POLYMER SCIENCE AND TECHNOLOGY
Volume 3

POLYMERS AND ECOLOGICAL PROBLEMS

POLYMER SCIENCE AND TECHNOLOGY

A Continuation Order Plan is available for this series. A continuation order will bring delivery of each new volume immediately upon publication. Volumes are billed only upon actual shipment. For further information please contact the publisher.

POLYMER SCIENCE AND TECHNOLOGY
Volume 3

POLYMERS AND ECOLOGICAL PROBLEMS

Edited by
James Guillet

Lash Miller Chemical Laboratories
Department of Chemistry
University of Toronto
Toronto, Canada

PLENUM PRESS · NEW YORK-LONDON · 1973

Proceedings of the Symposium on Polymers and Ecological Problems,
American Chemical Society, New York City, August 27-September 1, 1972.

Library of Congress Catalog Card Number 73-81406

ISBN-13: 978-1-4684-0873-7   e-ISBN-13: 978-1-4684-0871-3
DOI: 10.1007/978-1-4684-0871-3

© 1973 Plenum Press, New York
Softcover reprint of the hardcover 1st edition 1973
A Division of Plenum Publishing Corporation
227 West 17th Street, New York, N.Y. 10011

United Kingdom edition published by Plenum Press, London
A Division of Plenum Publishing Company, Ltd.
Davis House (4th Floor), 8 Scrubs Lane, Harlesden, London, NW10 6SE, England

To

O.V.L.

It is a characteristic of wisdom
not to do desperate things.

Henry D. Thoreau

PREFACE

   The growing public concern about environmental matters
has prompted widespread discussion in the media. Unfortunately
much of this public debate has been characterized more by ardour
than by information, and often the wildest speculations are pro-
mulgated with the same appearance of veracity as hard scientific
facts. It is an important, and often neglected, duty of scientific
societies to make sure that the public is properly informed regarding the technical aspects of matters of public interest, and to
assure that policy decisions of governmental and other agencies
are made with due regard to the scientific and technical facts, so
far as they are ascertainable.

   For a variety of reasons, not all of which are related to the
magnitude of the problems, a great deal of public attention has
been focused on the environmental aspects of the chemical industry.
Because of this the American Chemical Society has wisely decided
to sponsor a number of symposia at national scientific meetings
where these issues can be raised and information supplied regarding
their technical and scientific aspects.

   This book is the result of the collaboration of a variety of
individuals from many disciplines who contributed to the program
of the Symposium on Polymers and Ecological Problems held in
the Statler Hilton Hotel during the New York ACS meeting, August
28 and 29, 1972. The symposium itself was sponsored by three
divisions of the Society: the Polymer Division, the Organic Coatings and Plastics Division, and the Cellulose, Wood and Fiber
Division. As a representative of the Polymer Division I was asked
to coordinate the total program. In this task I was ably assisted
by the representatives of the other sponsoring bodies: Robert F.
Schwenker, Jr. of the Cellulose, Wood and Fiber Division, Bailey
Bennett of the Organic Coatings and Plastics Division, and William
J. Bailey of the Committee on Macromolecules of the National

Academy of Science - National Research Council. Although the job
of coordinating a meeting sponsored by so many bodies is not an
easy one, it was made enjoyable by the unfailing good humor with
which my often querulous requests for information were received.
Although the number of papers presented was not as great as we
had originally anticipated, it is my feeling that they made up in
quality what we missed in quantity. I am especially pleased that
all of the authors who presented papers in the symposium agreed
to prepare their material in textual form for this book and hence
this represents a more or less complete record of the proceedings of this important meeting.

I would also like to pay tribute to the chairmen of the three
sessions, Dave Wiles, Bob Schwenker and Bill Bailey, who did
much to enliven the discussion which followed the technical presentations. This is often a part of the program which is neglected,
but in this instance much useful information was brought out which
would otherwise have been missed.

I regret that the poor quality of the recording of the panel
discussion at the end of the symposium made it impossible to include an exact transcription. Many important issues and much
pertinent discussion originated from the floor, but only the answers
of the panel were audible on the tape. Because of the importance
of many of the issues dealt with by the panel I have included all of
the tape which could be transcribed in the book, and have para-
phrased the questions to give some continuity to the discussions.
I herewith tender my apologies to those who contributed to the discussion, and may not find their exact words in the text. My justi-
fication is that I felt that a partial record of the proceedings was
better than none at all.

As an author I would also like to acknowledge my personal
indebtedness to John W. Tamblyn and James Henry of the Tennessee
Eastman Company, whose early work in this field did much to stim-
ulate my own ideas. Mr. Oscar Van Leer of the Royal Packaging
Industries Van Leer N.V. of the Netherlands and Mr. Paul J.
Wright of EcoPlastics Limited of Toronto have done more than
anyone else to bring what was only a gleam in a scientist's eye to
the reality of commercial exploitation. Others, too numerous to

mention, both at the University of Toronto and in the English and Danish laboratories associated with the Van Leer Group, have made important contributions, for which I am only a spokesman.

This book would not have been possible without the work of Miss Susan Arbuckle. With patience, tact and skill she carried out the voluminous correspondence, typed the final manuscript, prepared the index and even drew the illustrations. Whatever success it may have is due almost entirely to her efforts.

James Guillet

Toronto, Canada
March 7, 1973

FOREWORD

The Symposium on Polymers and Ecological Problems sponsored jointly by three American Chemical Society Divisions, in cooperation with the National Academy of Sciences, was organized by Professor J. E. Guillet, Department of Chemistry, University of Toronto. Professor Guillet's program provided a valuable forum for the consideration of this very timely topic; for the first time, virtually all aspects of the plastics waste situation were discussed.

DEGRADABLE PLASTICS

Of great interest to polymer chemists, plastics technologists and applied microbiologists alike were the papers on photodegradable plastics for packaging and other single-use applications. Dr. Bernard Baum (De Bell & Richardson) described selected catalysts-- two and three component prodegradant additive systems --which could be used for polyolefins. A dark reaction was reported for at least one of the photoactivators in poly(propylene). Additional subtleties were emphasized by Professor Gerald Scott (University of Aston) in the description of his photo-accelerator additives which provide for a predictable induction period before, and advance warning of, the onset of physical disintegration of polyolefin plastics. Low levels of anti-oxidants, such as metal dialkyldithiocarbamates, provide transient protection initially but the degradation products of them subsequently catalyze the photochemical destruction of plastics.

In his lecture entitled "Polymers with Controlled Lifetimes", Professor Guillet described the significant merits of incorporating in the polymer molecules small amounts of vinyl ketone comonomers during resin synthesis. With the carbonyl-group carbon atom adjacent to (but not part of) the main polymer backbone, the sun's erythmal radiation causes embrittlement of these plastics over a wide range of predictable, and selectable, outdoor exposure periods. The long-term indoor stability and the processability of a range of possible copolymers is equivalent to that of the "parent" plastic; there are, of course, no problems of photosensitizer migration or loss.

Professors Scott and Guillet presented some evidence that the fine particles, resulting from the deterioration of their plastics by terrestrial sunlight, are microbially susceptible. There may be some question as to the suitability of certain accelerated fungal screening tests for providing evidence of rapid biodegradability, but the long-term (decades) assimilation of plastic fragments by fungi and/or bacteria is probably more relevant in any case. It does seem certain that most commercial and experimental polymers are not biodegradable, at least not prior to UV irradiation, according to Dr. R. A. Clendinning (Union Carbide). The large volume thermoplastics, owing to their molecular ordering and high molecular weight, seem to be resistant to micro-organisms even when "metabolically active" functional groups are attached to the polymer chains. A high molecular weight polyester has, however, been synthesized from ϵ-caprolactone and this plastic biodegrades in a soil burial test.

SOLID WASTE MANAGEMENT

The second part of the Symposium began with an eloquent overview by Dr. George Ingle (Monsanto) of the interaction of plastics and all aspects of the environment. In production, use, and disposal, the inherent properties of plastics can be considered to provide a "high benefit/risk ratio" from an environmental preservation point of view. G. L. Huffman (National Environmental Research Center) described methods for the disposal of plastics according to current technology in solid waste management. It was noted that discarded plastics need not be a particular problem in incinerator or sanitary landfill operations. Representatives of the Environmental Protection Agency provided opinions of some overall governmental views of how effects of solid waste on the environment could be minimized. John P. Lehman discussed how potential regulations could affect plastics recycling and other

resource recovery programs. Ralph Black emphasized the inade-
quacies of municipal waste disposal systems and indicated how
packaging manufacturers could help to cope with local problems
before these became crises.

COMBUSTION OR RECOVERY?

In the third part of the Symposium, some detailed consideration was given to the combustion propensity and other thermal
properties of thermoplastics. R. B. Engdahl (Battelle-Columbus)
described the incineration of plastics, especially methods for reducing the deleterious effects of HCl emission from PVC and
other chlorine-containing constituents of municipal waste. The
ignition and combustion properties of fiber assemblies was the
subject of a paper given by J. R. Martin (Textile Research Institute); the connection was shown between this basic research and
the development of improved incineration procedures for mixed
waste-fiber disposal.

Further consideration of PVC incineration was provided in
a lecture by R. Salovey (Hooker Chemical). He demonstrated
that pre-treatment of this plastic with ionizing radiation reduces
molecular order and facilitates subsequent thermal decomposition.
Finally, the unique ecological and economic incentives for the recycling of poly(tetrafluoroethylene) were outlined by Barry Arkles
(Liquid Nitrogen Processing Corporation). In this context he
described purification and depolymerization techniques which are
suitable for the recovery of the relatively expensive monomer.

DISCUSSION

Many aspects of the twelve lectures in the Symposium were
discussed and clarified in the Panel Discussion which followed the
more formal part of the program. Numerous questions from the
floor revealed uncertainties on the part of researchers, producers,
processors and fabricators -- at least some of whom expressed
as much concern for environmental preservation as for enlighten-
ment about the economics of proposed new technology. It was

clearly an advantage to have concomitant consideration of mutual
problems by experts from several disciplines (chemistry, engineering, microbiology) and representing several approaches (programmed degradability, incineration, government incentives and/or
regulations, recycling).

But what was settled? Is there a problem with waste plastics?
Are there viable solutions? At least for this observer, several
issues were brought into focus. Discarded plastics in general do
not appear to present especially difficult environmental problems,
but the use of plastics can provide uniquely advantageous solutions
in areas of current concern. Clearly, the science and much of
the technology have been elucidated for producing plastic packaging with a relatively short (and in certain cases well-controlled)
outdoor lifetime. This does seem an eminently reasonable approach
for coping with single-use thermoplastic materials which are not
amenable to regular collection/disposal procedures. Use of degradable plastics does not appear to create undue difficulties in
processing, or in conventional municipal waste disposal programs.
The common commercial thermoplastics are inherently resistant
to microbial disintegration but perhaps become biodegradable after
extensive photo-oxidation (it is not clear to me why this should be
considered essential, except for long-term preservation of the
carbon cycle).

Governments have a role to play in waste disposal programs
and, although plastics manufacturers or fabricators should not be
singled out for criticism, they can help by encouraging development of improved municipal systems. Solutions for apparent
problems with incineration of plastics are available or can be
readily developed. Short of official regulations and/or incentives,
large scale recycling of high volume used plastics does not appear
to be imminent; and the complex questions of resource/energy
conservation await resolution.

Undoubtedly, some symposium attendees will have reached
other conclusions or identified a different pattern of emphasis.
Nevertheless, it has surely been a useful exercise in collaboration and communication to examine many inter-related aspects

of "Plastics and Ecological Problems". The importance of the subject matter amply justifies the collection of the symposium proceedings in the present volume for dissemination to a wider audience.

D. M. Wiles
National Research Council
of Canada

Ottawa, Canada
November, 1972

EDITOR'S INTRODUCTION

Much of the weight of public concern about environmental matters has centered around activities of the chemical industry. The reasons for this are partly historical, since the industry is almost completely a product of Twentieth Century technology, whereas others such as mining and metallurgy, textiles, steel and coal are rooted more firmly in the past. There are today a number of people who follow the Luddite tradition and wish to blame all of society's ills on the development of modern technology. Because of this, the chemical industry has perhaps been subjected to more than its proper share of criticism. While it seems unlikely that philosophical concepts which would require individual citizens to forgo the obvious material benefits of technology will find general acceptance in democratic societies, this should not blind us to the fact that there are very serious issues at stake. These relate to problems such as the harmful effects of industrial pollution, the conservation and allocation of both energy and material resources and the long-term effects of modern technology on human health and the environment. Also it seems clear that in our haste to improve the material lot of mankind we may have neglected aspects which relate more to the aesthetic considerations frequently referred to as "the quality of life".

A very large part of the chemical industry is concerned with the synthesis and use of polymeric compounds usually derived from petrochemical raw materials. These include synthetic fibers and plastics, paints, adhesives and other resinous products. This book is the result of an attempt to identify and discuss certain problems relating to polymeric materials and their impact on the environment.

Environmental issues relating to the chemical industry can be divided into three broad categories:

1. Pollution relating to improper control of effluents from chemical manufacturing plants;

2. Consideration of the conservation or allocation of mate-
    rial or energy resources in the manufacture of chemical
    products; and

3. Problems relating to the use of chemical products in
    commerce, household or industry.

This book is concerned primarily with the latter issue as it relates
to polymeric materials generally and more particularly to plastics
and synthetic fibers.

Although there are a variety of topics which surely merit
consideration under the general title "Polymers and Ecological
Problems", in recent years public concern has mainly centered
around a single issue, namely the disposal of plastic waste. A
Freudian psychologist might be inclined to categorize the entire
matter as being the result of an "anal fixation" on the part of poli-
ticians, press and the public. Be that as it may, all of the papers
presented at the symposium dealt almost exclusively with various
aspects of this issue.

Synthetic polymers (plastics) are now used extensively in
modern packages. Because they are light, inexpensive and have
a remarkable combination of physical properties they represent
a small but rapidly growing proportion of packages used, particularly for food and beverages. Because of their low cost they have
been used extensively for so-called "disposable" or "throw-away"
packaging and are often inseparably associated with this use in the
public mind. Because of their unique combination of properties,
plastics have begun to replace more traditional packaging materials
based on cellulose, glass or metal, and much of the disparaging
publicity regarding the use of plastics has come from these other
industries seeking to retain their share of a lucrative market.

The major complaints which can be identified at the present
time are:

1. Problems relating to the disposal of plastic and fiber
    waste in garbage by (a) sanitary landfill, (b) incineration,
    and (c) garbage dumps;

2. Problems relating to the longevity of plastic packaging when it eludes garbage collection systems and becomes permanent litter;

3. Problems relating to recycling or reuse of plastics recovered from scrap or garbage; and

4. Problems relating to allocation of resources and energy, i.e., given that certain of our natural resources are nonrenewable and that energy supplies are limited, what considerations other than price should govern the selection of a package between say cellulose, plastic, metal or glass?

All of these issues, with the possible exception of the last, are treated in some detail in various sections of the book and it would be inappropriate for me to try to summarize the conclusions. There are, however, two matters which do deserve comment because they are not dealt with specifically and they are subject to popular misconceptions.

The first of these misconceptions is that plastics are unnatural materials whose structure is completely different from anything ever present on the earth before. The argument is that such materials synthesized artificially by man are inherently incompatible with the natural environment and must therefore lead to long term ecological damage. Proponents of this view must ascribe to the ancient philosophical concept of the "vital force" present in chemicals manufactured by Nature which is supposedly absent in materials synthesized by man. The death blow to this theory was given by Wohler's synthesis of urea in 1828 but it apparently has lived on in the minds of some of the eco-activists.

In fact, modern synthetic plastics are usually composed of the same elements, and in rather similar arrangements, as biological polymers such as cellulose, collagen and keratin (wood, tendon, hair). They are truly organic compounds and as such are more likely to be compatible with living systems than inorganic materials such as metal or glass. Possibly some of the concern about the "biodegradability" of plastics is related to these considerations. However as is clearly demonstrated in the proceedings of the symposium, it is possible to make completely synthetic plastic

materials which are fully biodegradable should this indeed prove
to be a desirable characteristic.

The second issue is the relationship of plastics and synthetic
fibers to the depletion of our non-renewable resources. The argument is that since plastics and synthetic fibers are based on petroleum raw materials we should discourage their increased use in
order to conserve our scarce reserves of petroleum. Furthermore
it is suggested that since all petroleum reserves are going to be
used up in 20 to 30 years, we will be forced to find alternative
materials for future applications. Such arguments cannot be supported by the facts. At the present time only about 2% of all
petroleum and natural gas production is used as a source of chemical raw materials, the remaining 98% being used exclusively for
energy production. It is clear that it is a most urgent problem to
find alternative sources of energy, and that if we fail to do this in
time the "Doomsday" predictions of such bodies as the Club of
Rome will indeed be realized. On the other hand, even if we were
to eliminate completely all chemical uses of petroleum it would
have only a marginal effect on the energy crisis.

The basic raw materials for organic polymers are carbon,
hydrogen, oxygen, nitrogen and chlorine. All of these elements
are in plentiful supply in the air or the sea. The economics of the
chemical industry are based on the use of the cheapest source of
carbon. The ancient industry was based on carbon from charcoal
which came from the pyrolysis of wood. With the discovery and
exploitation of coal, in the eighteenth and nineteenth centuries,
this became the primary carbon source of the chemical industry.
We still speak of "coal tar dyes" even though these are made almost
exclusively from petroleum sources at the present time. The development of the petrochemical industry began in the early part of
the present century but it did not become a major factor in chemical production until after the Second World War. If and when present reserves of petroleum or natural gas are completely depleted,
it is clear that the chemical industry will continue to operate, but
using alternative sources of carbon. Since most predictions suggest that coal reserves will outlast petroleum, perhaps by as much
as a century, it seems logical to predict that in the future coal will
again become the basic raw material. However vast reserves of

carbon are also available in the carbon dioxide in the air or dissolved in the sea, in carbonate locked in limestone deposits, and
in plant sources such as cellulose. In one sense carbon is a truly
renewable resource since the carbon dioxide in the air is being
continually concentrated by photosynthesis in plants as part of the
carbon cycle of the natural environment. Because of this renewable aspect of carbon supplies for chemical uses one could make a
strong case for the replacement of many other materials which do
deplete scarce resources with plastics and other synthetic polymers.

The polymer industry has often tended to react defensively
to many of the issues discussed in this book. In my view this is
a mistake, since a full discussion can only lead to a better understanding, both by scientists and the public, of the unique role played
by these remarkable materials in our society. It is my hope that
this volume will contribute to such an understanding.

J. G.

CONTENTS

LIST OF CONTRIBUTORS

ACKART, W. B.	KRAUSE, H. H.
ARKLES, B.	LEHMAN, J. P.
BADGER, R. G.	MARTIN, J. R.
BAUM, B.	MEISER, C. H. Jr.
BLACK, R. J.	MILLER, B.
CLENDINNING, R. A.	MILLER, P. D.
ENGDAHL, R. B.	NIEGISCH, W. D.
GUILLET, J. E.	POTTS, J. E.
HUFFMAN, G. L.	SALOVEY, R.
INGLE, G. W.	SCOTT, G.
KELLER, D. J.	WHITE, R. A.

POLYMERS WITH CONTROLLED LIFETIMES

J. E. Guillet

Department of Chemistry
University of Toronto
Toronto, Canada

INTRODUCTION

It has become popular in certain circles to blame many of
our contemporary problems on advances in technology. In keeping with this trend, we often blame the packaging industry for
increasing amounts of garbage and litter, and the disposal problems
associated with it. The fact that the development of the modern
packaging industry has had a major effect on reducing the cost of
consumer goods both in manufacture and distribution, and the critical
role it plays in protecting our health is often ignored. However,
it must in fairness be admitted that until very recently the packaging industry has been so concerned with the problem of providing
adequate and attractive packaging that it neglected to concern itself
about the disposal of these packages once they had served their
initial functions. Ecologists have made us fully aware that no
human system can be considered in isolation. Each human activity must be contemplated in relation to our total environment. We
should therefore extend our consideration of packaging practice to
the point of ultimate disposal or recycling of the material from
which the package is made.

At this point it is helpful to distinguish between two kinds of
disposal problems relevant to our modern society. The first of
these is garbage disposal. Garbage can be defined as the discarded

1

solid waste products of household or industry which are <u>collected and disposed</u> of in some central facility such as a dump, land fill, or incinerator. Litter, on the other hand, may be defined as a man-made object in a place where it should not be. For example a fallen tree in the forest is not litter, but a discarded wooden box made from the same material in the same place is. Paper in a garbage can is not litter. The same piece of paper blowing along the side of a road definitely is. Surveys of litter show that by far the greatest proportion is containers or packages of various kinds used for food, beverages or tobacco.

The present symposium is concerned with the role that plastics play in both of these problems, in particular in the latter. Plastics pose some special problems with respect to garbage disposal, but they still represent a rather small portion (less than 10%) of total municipal garbage and it is our contention that those problems that do exist can be solved by appropriate engineering and design of the disposal facilities. Until these problems have been more clearly defined, it does not seem appropriate at this point to suggest an alteration in the properties of plastics which would make them more suitable for garbage disposal. Plastics litter, on the other hand, does pose special problems because these new synthetic materials resist natural degradation processes, and hence form a kind of "permanent" litter. This is not a property that is inherent in all plastic materials, and we have shown that it can be altered at will by the use of appropriate technology.

Containers have been used from the earliest beginnings of civilized man and in fact one could make a cogent argument that development of containers was essential for the development of civilization. It certainly was vital that man became able to store and carry food in order to develop any civilized community and to extend his knowledge and contacts with other groups of his own kind. The only modern development might be considered to be "disposable" packaging. In early times, the effort required to make a package such as a wine jar or a basket was such that it was customary to use the same package many times. Paper, of course, although it was discovered thousands of years ago, was extremely expensive and was not used as a temporary package until the development of large scale paper-making machinery in the last century.

However, much as we would like to attribute the disposable package to man's genius, it is clear that nature had anticipated the development by millions of years. It is not difficult to list a large number of edible products produced and packaged by nature in disposable packages. Typical examples are (1) a coconut in a coconut shell, (2) a banana in a banana skin and (3) a clam in a clam shell. Although these packages are made of widely different materials, they have one characteristic in common and that is that they are naturally recycled into other useful products by natural processes. The first two examples degrade biologically over a period of time, the banana skin rather rapidly and the coconut shell rather slowly, to become part of the natural soil and nourish further growth through decomposition by micro-organisms. A clam shell on the other hand breaks up under the action of mechanical forces to become sand, and if a sufficient number are accumulated in one place, may ultimately turn to limestone.

It is this latter feature of natural packages which so far has distinguished them from the synthetic products produced by man. A glass bottle represents a more or less permanent package unless it is broken, in which case it becomes an even more undesirable form of litter. No one knows how long a discarded aluminum can will last in a natural environment, but one would guess it would be at least a century. Steel cans will last over a period of five to ten years although this period has been extended by the modern practice of protecting them with plastic protective coatings. Similarly, plastics litter (as currently manufactured) can be expected to last several years at the minimum and may well last considerably longer. Needless to say, the longer the lifetime of the package, the greater will be its tendency to accumulate in the less accessible areas of the countryside, where it is not possible or desirable to send in crews of people to pick up discarded litter.

It is imperative that a solution to this problem be found, because litter is not a problem that concerns only the civilized or heavily populated portions of the world. On a deserted beach in the Caribbean one can find literally hundreds of aluminum cans, plastic bottles, drinking cups, etc. These are not discarded by the local population, but are carried by the water from populated areas such as Florida, or from ocean-going cruise ships. The

problem is not confined to deserted beaches. In the arctic regions
an accumulation of litter discarded over many years can be found
near every settlement. There the problem is more acute because
the very low temperatures which prevail inhibit almost all biological activity through which natural biodegradation might occur.

Wherever we go we find litter, and often instead of blaming
the person who discarded it, we blame the manufacturer. Aluminum beverage cans are permanently marked with the name of
the product and its manufacturer. They do not have a paper label
that will wash or tear away. Plastic bottles often have the name
of the product permanently molded into the container. This kind
of "negative advertising" has put manufacturers under increasing
pressure to use packages made of some sort of degradable material.

Plastics have one major advantage over glass and metal in
this respect and that is that they are inherently organic materials
just like the banana skin and the coconut shell and it is therefore
possible in principle to make them degrade by natural biological
mechanisms once they have performed their primary function as a
temporary container. This paper describes a new way in which
this can be accomplished.

In consideration of these principles, it is possible to draw
up a list of the desirable characteristics for a packaging material.

1. It must be resistant to the material which it is to contain and
   not contribute to the taste, odor, or toxicity, particularly if
   it is a food product.
2. It must be light in weight and easily formable into an attractive package.
3. It must be cheap and represent a minimal expenditure of
   natural nonrenewable resources in its manufacture.
4. It must be resistant to micro-organisms which might otherwise attack the materials which it contains.
5. It must be stable and maintain its desirable physical properties for at least the lifetime of the product which it contains.
6. It must be disposable or recyclable by conventional garbage
   disposal technology.
7. It should degrade by some natural mechanism if it becomes
   litter.

Plastics as currently manufactured, fulfill all of these characteristics except the last, and until recently the latter requirement has been considered to be inconsistent with the fourth since it was felt that if the plastic were biologically degradable, it would no longer afford adequate protection against the attack of micro-organisms on the product which the plastic is intended to contain. Now, however, as a result of research carried on at the University of Toronto over the past decade and in other laboratories throughout the world, it is clear that these two requirements need not be mutually exclusive. It has been found that the resistance of conventional plastics to micro-organisms is primarily due to two factors: (1) the low surface area and relative impermeability of plastic films and molded objects and (2) the very high molecular weight of the plastic material. Micro-organisms tend to attack the ends of large molecules and the number of ends is inversely proportional to the molecular weight. In order to make plastics degradable, it is necessary first to break them down into very small particles with large surface area, and secondly to reduce their molecular weight.

We have found that this can be done utilizing the energy of natural sunlight. The principle of the method is outlined hereafter.

TECHNOLOGY OF DEGRADABLE PLASTICS

Recently a number of processes have been disclosed using additives to make "unstable" plastics. But packaging materials require not only a polymer that will degrade, but one that will degrade at a controlled and predictable rate. Our approach to this differs in that we make a change in the plastic molecule itself. Small quantities of a sensitizing group are chemically attached to the macromolecular chains. When a plastic containing this sensitizing group is exposed to natural sunlight, the sensitizing group absorbs radiation which causes the chain to break at that point and thus form smaller segments. Since the physical properties of a plastic depend on the length of the chain, if the chain is broken, it will become very fragile. If it is broken in enough places, it becomes biologically degradable.

The process developed after many years of research in the photochemistry of polymers at the University of Toronto involves inclusion into the backbone of the chain of a polymer a group of the general structure

$$
\begin{array}{c}
R \\
| \\
\text{\textasciitilde} - C - \text{\textasciitilde} \\
| \\
C{=}O \\
| \\
R'
\end{array}
$$

where R and R' are various alkyl and aryl substituents. The reason for the substituents is that by changing the nature of these two groups, one can control the rate of the degradation process. When the carbonyl group absorbs a quantum of ultraviolet light, an electron localized on the oxygen atom is raised to the π^* state in which the electron is delocalized over the C$-$O bond in an anti-bonding molecular orbital. This is the lowest energy transition of most carbonyl compounds. From this excited state, the energy must be released and there is a good probability that this will be done by a photochemical reaction.

The classical photochemical reactions which occur are (a) the Norrish type I reaction, a free radical split occurring at the carbonyl group to give two free radicals, and (b) the type II process, an intermolecular rearrangement resulting in a scission of the main chain to give a methyl ketone and a terminal double bond.

The type I process gives two free radicals, which will not, in the absence of oxygen, result in a break in the chain. The type II reaction is the major photodecomposition process that causes the chain to break. However, in the presence of oxygen, the radical sites can introduce photooxidation processes which cause chain degradation over a longer time scale.

Another characteristic of these two processes is that they have quite different temperature dependences as illustrated in a typical Arrhenius plot (Fig. 1). [1] There is a factor of about 10 difference between the quantum yields of the type II and type I reactions, even at relatively high temperatures. At ambient temperatures, however, there is a 100-fold difference in the quantum yield and the type II reaction is the only process which is of real significance as far as the chain breaking process is concerned. The other factor of importance is that this process is almost independent of temperature. The activation energy is less than 1 kcal/mole whereas the free radical process has an activation energy of about 5 to 6 kcal mole^{-1}. Photooxidations catalyzed by additives such as benzophenone also seem to have similar or higher temperature coefficients.

Figure 2 shows the absorption and emission spectra of the ketone carbonyl in typical ketone polymers along with those of acetone and 2-heptanone. [2] All the ketones have absorbances with a maximum of around 280 - 290 nm and they cut off rather sharply at about 330 nm. This is rather important since most synthetic polymers do not absorb in the region above 300 nm. Typical absorption spectra are shown in Figure 3. Figure 4 shows the solar spectrum along with that of other light sources. The emission spectrum of the sun has an approximate Boltzmann distribution of radiation which cuts off rather sharply at 300 nm because of the absorption in the ozone layer in the upper atmosphere. On the other hand, the emission of a typical fluorescent lamp and an incandescent lamp cut off rather sharply around 330 nm. As a result, ketones are rather stable compounds photochemically as far as visible light is concerned and they only undergo photochemical reactions if they are irradiated with light of wavelength less than 330 nm, such as occurs in natural sunlight. This gives the possibility of producing a packaging material which is stable in visible light, but which degrades when thrown away in an outdoor environment. This critical portion of the sun's spectrum, between

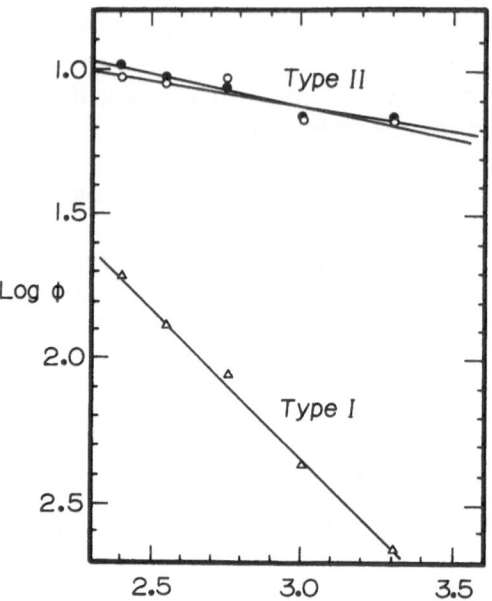

Figure 1. Arrhenius plots for type I and type II reactions in 8-pentadecanone.

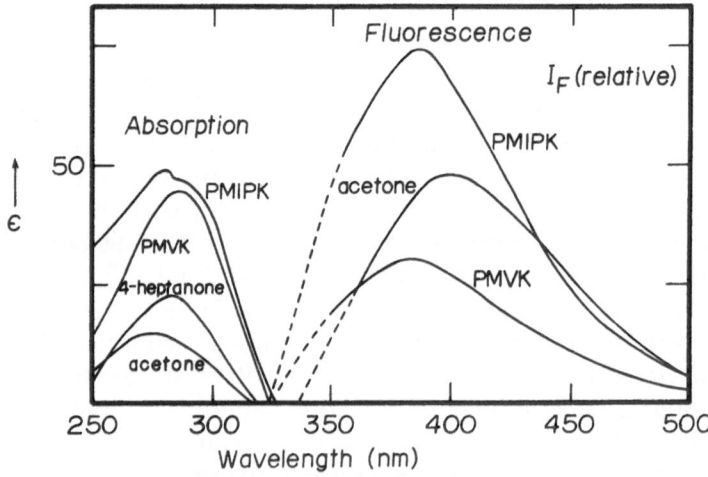

Figure 2. Absorption and fluorescence spectra of polyketones.

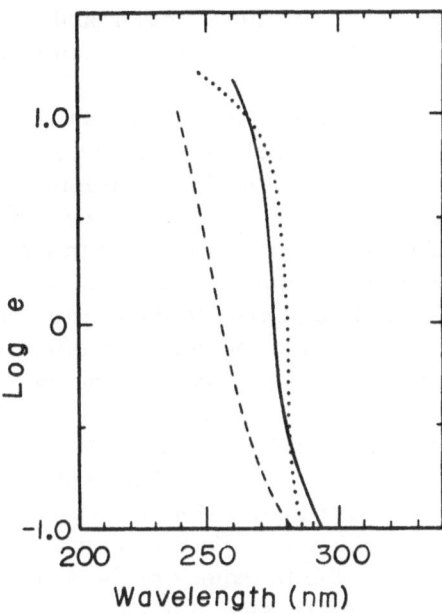

Figure 3. Absorption spectra of various polymers: – – – – poly-(methyl methacrylate); ———— poly(styrene); · · · · poly(vinyl acetate).

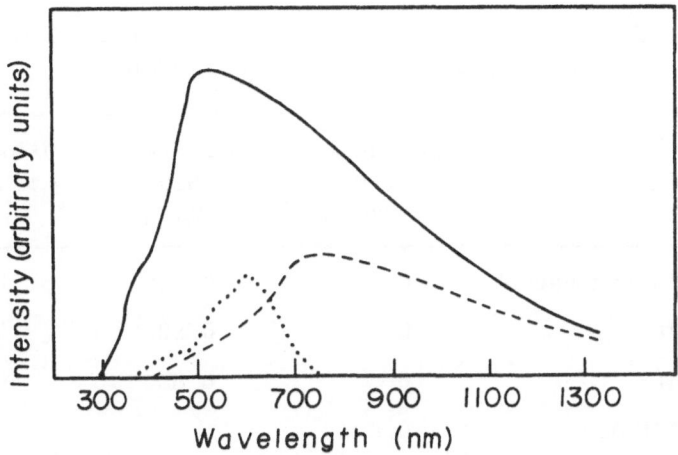

Figure 4. Wavelength distribution of light sources: ———— the sun; – – – – incandescent lamp; · · · · fluorescent lamp.

290 and 330 nm is called the erythmal region and is the radiation
responsible for tanning and sunburn of the human skin.

Table I shows the relative intensity of sunlight in various
regions of the spectrum. For the erythmal radiation, noon sunlight in Arizona is about 300 or 400 times the intensity of an
ordinary fluorescent light. This means that you would have to sit
a long time under a fluorescent lamp in order to get much of a tan.
In the near UV range there is more radiation in artificial sources
although still rather small compared to solar radiation. It is
quite obvious that there will be a much larger effect in sunlight
than there is under any of these normal lighting conditions.

The other factor which is important is that ordinary window
glass filters out the erythmal radiation of the sun. Figure 5 shows
that there is practically no transmission of the radiation in the
wavelength range 300 to 320 nm. Above 320 nm there is a slight
transmission depending on the thickness of the window glass and
above 330 nm an appreciable transmission occurs. So if one wants
a material that will be stable behind window glass, it should not
absorb radiation above 330 nm.

TABLE I

OUTPUT OF ARTIFICIAL LIGHTING COMPARED
TO SOLAR RADIATION (μwatts/cm^2)

Type of lamp	Erythmal $\lambda = 280 -$ 320 nm	Near UV $\lambda = 320 -$ 400 nm	Visible & IR $\lambda > 400$ nm
40 watt incandescent	0	0.21	21
100 watt	0	0.89	71
500 watt	0	6.55	409
40 watt fluorescent	0.8		
Noon sunlight, Arizona	259	4640	88,000

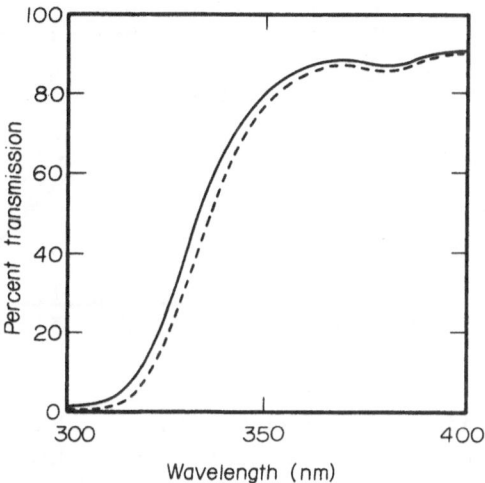

Figure 5. Transmission for two thicknesses of window glass:
———— 0.095" thick; — — — — 0.121" thick. (Data supplied
courtesy of Pittsburgh Plate Glass Co.)

It is not necessary for the plastic to be in direct sunlight in
order for the degradation process to occur. As shown in Figure 6,
over 50% of the total amount of ultraviolet radiation comes from
the sky rather than from the sun itself. Consequently even if the
plastic is in the shade it will still be receiving skylight and hence
will degrade. In fact, as long as the plastic can be seen outdoors,
it will be undergoing degradation. The rate at which the chains
will be broken depends only upon the intensity of the ultraviolet
light absorbed by the sample. In northern latitudes such as Canada
the intensity of this ultraviolet light will vary with the time of year
as shown in Figure 7. This means that in the winter time the rate
of degradation will be rather slow, while in the summer time, the
rate will be considerably more rapid. In equatorial regions the
intensity of UV radiation does not vary appreciably throughout the
year.

Rather surprisingly, the total amount of UV radiation does
not vary much over the surface of the globe. During the arctic
summer, for example, the UV radiation quantity is comparable to
that of more temperate regions simply because of the longer day-
light hours. This then provides a mechanism for degradation in
very cold arctic regions where biological processes are either very
slow or non existent.

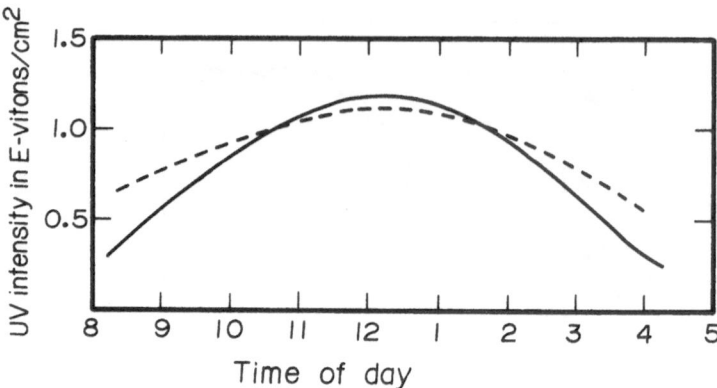

Figure 6. Intensity of ultraviolet radiation on a horizontal surface from the sun and from the sky at various hours throughout the day: ———— from the sun; – – – – from the sky. (From M. Luckiesh, Germicidal, Erythemal and Infrared Energy, Van Nostrand, New York, 1946.)

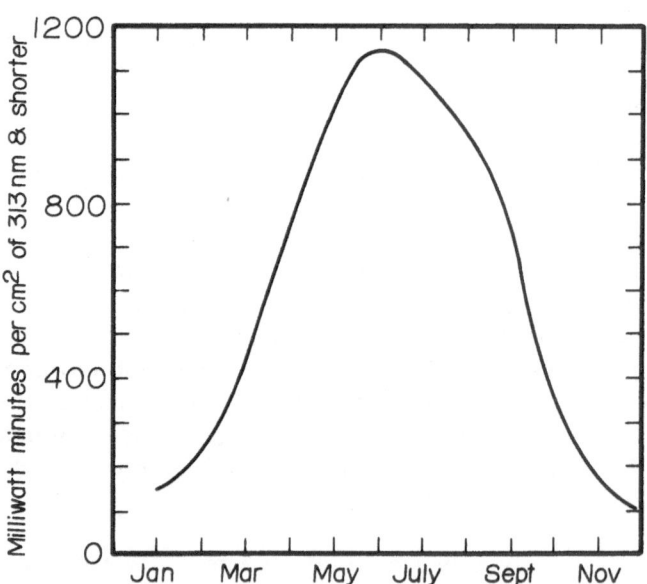

Figure 7. Integrated monthly ultraviolet of wavelength less than 313 nm throughout the year. (From W. W. Coblentz and R. Stair, J. Res. Natl. Bur. Standards, 33, 21 (1944)).

Rates of Photochemical Reaction

When dealing with photochemical reaction rates it is usual
to discuss them in terms of quantum yields. The quantum yield
for any process is the number of chemical events of a particular
kind which occur, on the average, per quantum of light absorbed.
For a non-chain process the maximum value of the quantum yield
is unity, and normally the value will be less than that. If we know
the average intensity of daytime sunlight, we can predict the time
required for a photodegradation process for plastic specimens of
different thickness. Table II shows such a prediction, assuming
that one needs ten chain breaks per polymer molecule, for different
quantum yields. This is really just an approximate guide to show
what sort of yields are required. It is quite clear that quantum
yields of 0.001 or higher are needed in order to get practical rates
of photodegradation even of thin specimens. Quantum yields in
excess of about 0.01 are required for a photochemical process for
polymer degradation in thicker samples.

Table III shows how the quantum yields depend on chain length
and the structure of the ketone carbonyl group.[3]   For a methyl or
phenyl ketone, the quantum yield remains almost independent of
chain length whether it is in a polymer or a small molecule.
Whereas if you include it in the center of a chain (this would be
equivalent to an ethylene carbon monoxide polymer), the quantum
yield decreases with chain length. However a side chain ketone
shows a considerably higher efficiency which is more or less
independent of molecular weight.

Figure 8 shows typical photolysis data from which quantum
yields ϕ_{cs} for chain scission may be determined for a copolymer
of methyl methacrylate and methyl vinyl ketone in solution.[4]
Methyl vinyl ketone is a very convenient way of introducing these
ketone carbonyls into a polymer structure. A plot of $(M_v^0/M_v) - 1$
as a function of light absorbed gives a straight line whose slope
is proportional to ϕ_{cs}. This expression represents the number
of bonds broken per original polymer molecule.

Of course, photochemistry changes in the solid phase, and
the quantum yields are affected very much by the mobility of

TABLE II

TIME REQUIRED FOR TEN CHAIN BREAKS
PER POLYMER MOLECULE[*]

Quantum yield	Thickness				
	100 mil	50 mil	10 mil	5 mil	1 mil
ϕ = 1.0	12 days	6 days	1 day	12 hrs	2 hrs
ϕ = 0.2	2 mos	1 mo	6 days	3 days	1 day
ϕ = 0.1	4 mos	2 mos	12 days	6 days	1 day
ϕ = 0.02	20 mos	10 mos	2 mos	1 mo	6 days
ϕ = 0.01	40 mos	20 mos	4 mos	2 mos	12 days
ϕ = 0.001	34 yrs	17 yrs	3.2 yrs	1.6 yrs	3 mos
ϕ = 0.0001					30 mos

*Based on UV data, Washington, D. C., average.

TABLE III

VARIATION IN ϕ_{II} WITH CHAIN LENGTH

Chain length R (carbon atoms)	$\overset{\text{O}}{\overset{\|}{R-C-R}}$	$\overset{\text{O}}{\overset{\|}{R-C-CH_3}}$	$\overset{\text{O}}{\overset{\|}{R-C-\bigcirc}}$
4	0.11	0.25	0.31
5		0.20	0.31
6	0.092	0.20	0.25
7	0.080	0.20	0.30
8			0.29
9		0.20	
11	0.072		0.31
17		0.15	
21	0.059		
100*	0.025	~0.2	0.2-0.3

* Estimated value for polymer with isolated ketone chromophore from experimental data on ketone copolymers

molecules in the solid state. Figure 9 shows the variation of ϕ_{cs}
with temperature when the polymer is irradiated as a solid film
for a copolymer of styrene and phenyl vinyl ketone.[5] A quantum
yield of about 0.04 is obtained at 20° and this increases slightly
up to the glass transition T_g. A remarkable increase in ϕ_{cs}
occurs at T_g and above this the value is constant and equal to that
obtained in solution. This implies that the mobility involved at the
glass transition is very similar to that in solution and results in
similar quantum yields for photolysis. However the values even
below T_g are well within the range of quantum yields that are useful for photodegradation. Similar results are shown in Figure 10
for a MMA-MVK copolymer.

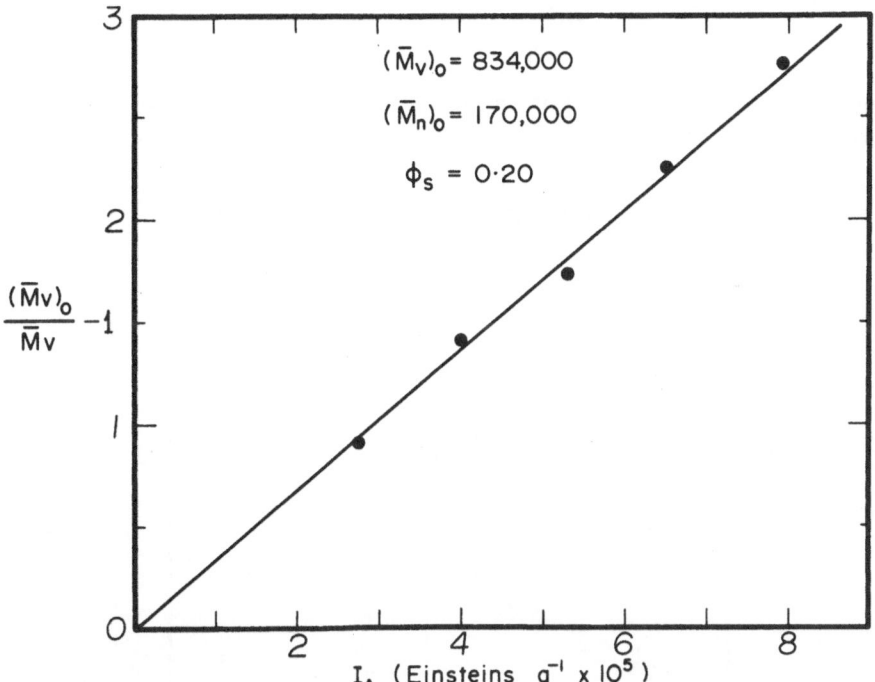

Figure 8. Photodegradation rate for high molecular weight MMA-
MVK copolymer containing 2.9% MVK, at 313 nm in benzene solution at 25°, determined by the single-point viscosity method.

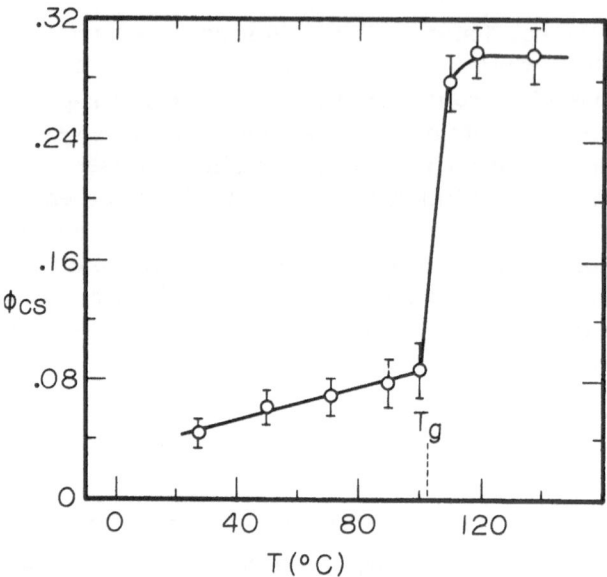

Figure 9. Quantum yield of chain scission as a function of temperature, PS-PVK film.

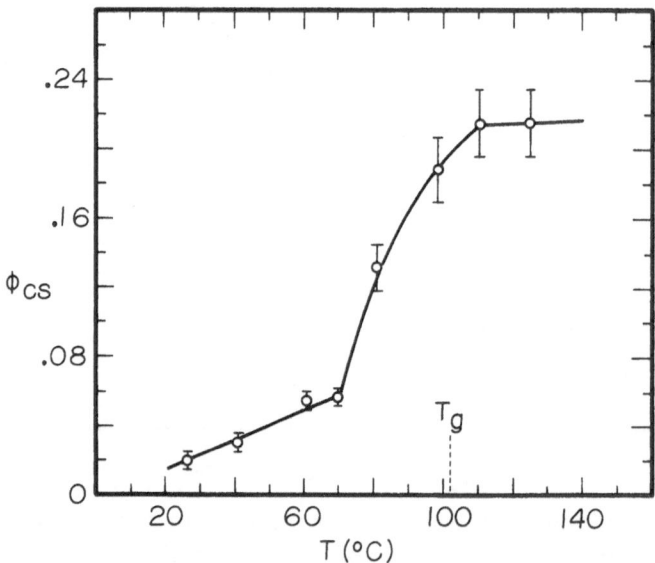

Figure 10. Quantum yield of chain scission as a function of temperature, PMMA-MVK film.

THE DEVELOPMENT OF ECOLYTE* PLASTICS

Using these basic principles it was found possible to develop photodegradable plastic compositions based on most of the large volume industrial plastics. These compositions and processes have been patented by the University of Toronto in over thirty countries of the world. The commercial development of these has been undertaken as a joint enterprise by two companies, Eco-Plastics Limited of Toronto, Canada, and Royal Packaging Industries Van Leer of Amstelveen in the Netherlands under license from the University of Toronto.

After extensive pilot plant and commercial evaluation, the first of these plastics, ECOLYTE S, has reached the production stage. This is a modified polystyrene suitable for the manufacture of disposable packaging of all types. ECOLYTE E, a modified polyethylene is also undergoing commercial and plant trials.

The basis of the process by which ECOLYTE plastic resins are made is to introduce into the structure of the polymer a sensitizing group which absorbs radiation in the erythmal region of the solar spectrum and which causes the chain to break at a point adjacent to the absorbing group. The chromophore is selected so that it does not absorb in the visible region of the spectrum, so the plastic sample remains stable in visible light, but will degrade when exposed to solar radiation.

The physical properties of any plastic depend on the length of the molecular chains. The greater the length of these chains, the stronger and tougher the plastics will be. As soon as the chains are broken, however, the plastic begins to become fragile, and if the process proceeds to a sufficient extent, the plastic becomes so brittle that it breaks up under the action of such erosive forces as wind, rain, and waves into very small particles. As the molecular chains are being broken, the molecular weight is reduced and the number of chain ends increases and the plastic now becomes more susceptible to biological degradation by micro-organisms

*Trade mark.

Although the chain breaking process begins as soon as the plastic is exposed to solar radiation, there is a certain period of time necessary before an appreciable change in the physical properties occurs. This is illustrated in Figure 11. The reason for this is that above a certain molecular weight, which is sometimes called the critical molecular weight, there is only a small change in the physical properties of the polymer as the molecular weight changes. Once this molecular weight is reached however, any subsequent decrease in molecular weight will cause a drastic change in the properties of the material. This means that even after exposure to solar radiation, the plastic material will still retain its useful properties for a certain period of time, and this time can be controlled at will in the manufacturing process.

The key to the process is the selection of the correct photo sensitive group to include in a particular plastic in order to give the desired rate of degradation. One of the chemical groups having the desirable characteristics necessary for this process is a ketone group. Ketones in general have the desirable characteristics that they absorb the ultraviolet light of solar radiation, but do not absorb visible radiation. Consequently they do not contribute to color, nor does the degradation process occur by the action of visible light.

Depending on the structure of the ketone group, and its location along the polymer chain, the efficiency of its reaction to ultraviolet light can be changed. Exhaustive studies of these systems in a variety of polymers have shown that the rates as measured by the quantum yields are independent of oxygen, moisture, and other experimental variables such as temperature and pressure, and consequently the rate can be accurately predicted from the intensity of the ultraviolet light and the time of exposure only. It is found that in order to provide an acceptable rate of degradation for polystyrene, for example, it is only necessary to include less than 1% of these carbonyl groups in the polystyrene molecule.

The rate of degradation is directly proportional to the concentration of carbonyl groups for a given thickness of a plastic specimen as shown in Figure 12. The rate is not much affected

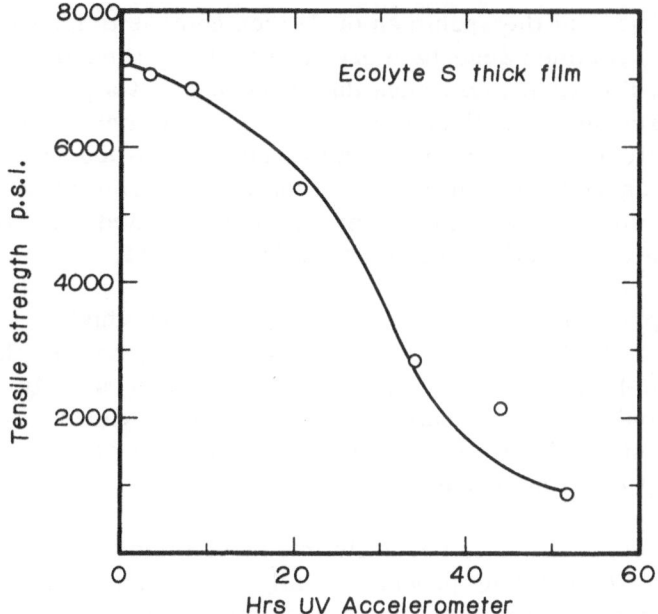

FIGURE 11. Variation in tensile strength with UV exposure for ECOLYTE S films (0.6 mm).

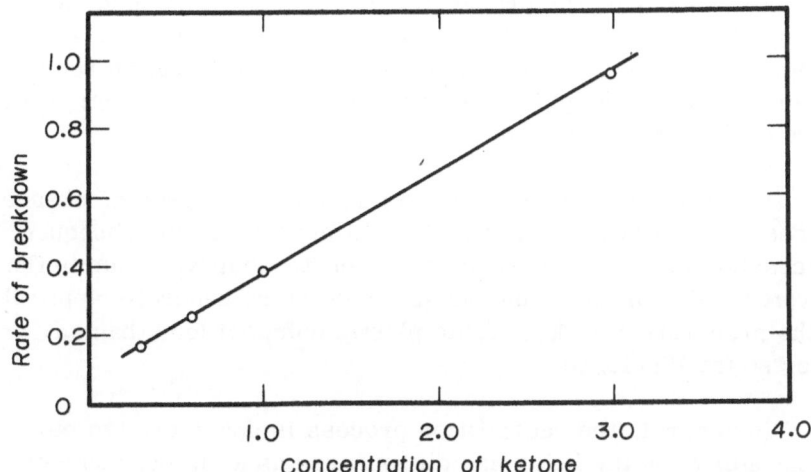

FIGURE 12. Rate of photodegradation as a function of ketone group concentration for ECOLYTE S films.

by the thickness of the specimen of a given composition because
the thicker specimen absorbs more light than the thinner. This
is shown in Figure 13. Because the process is not a photooxidation, the presence of anti-oxidants and other stabilizers does not
normally affect the rate of degradation unless they absorb in the
ultraviolet range which initiates degradation. Conventional
weathering stabilizers can however be used to retard the rate of
degradation if desired. This is shown in Figure 14.

The process works with a wide variety of vinyl
polymers, including poly(styrene), poly(ethylene), poly(methyl
methacrylate), poly(acrylonitrile) and its copolymers, poly-
(methacrylonitrile) and copolymers, and poly(propylene). In each
case, a substantial rate of degradation is achieved with minor
amounts of the ketone group.

The ketone group is normally introduced into the polymer
during the manufacturing process usually by copolymerization
with another monomer containing the group in the appropriate
position. With vinyl monomers this is usually done by copolymerization with a vinyl ketone monomer. In condensation polymers
such as polyamides and polyesters, the group is introduced by
synthesizing a difunctional monomer containing the ketone group
and adding it during the preparation of the polymer. In some
cases, ketone groups can be introduced by a chemical post-
treatment. Because of the minor amount of modification required,
the properties of the photosensitive resin are almost identical with
those of the untreated plastic.

In all of the cases studied so far, there has been no change
necessary in either the polymerization process or in subsequent
processing due to the presence of the photosensitive group in the
polymer. The process thus seems to be more generally applicable
to the preparation of degradable plastic compositions than any
other so far disclosed.

In general, the degradation process is considered to be
successful when the litter has disappeared as a visible man-made
object. This means that at this point a package has broken up
into small particles comparable to sand and is dispersed in the
surrounding soil. The rate of this process can be accurately

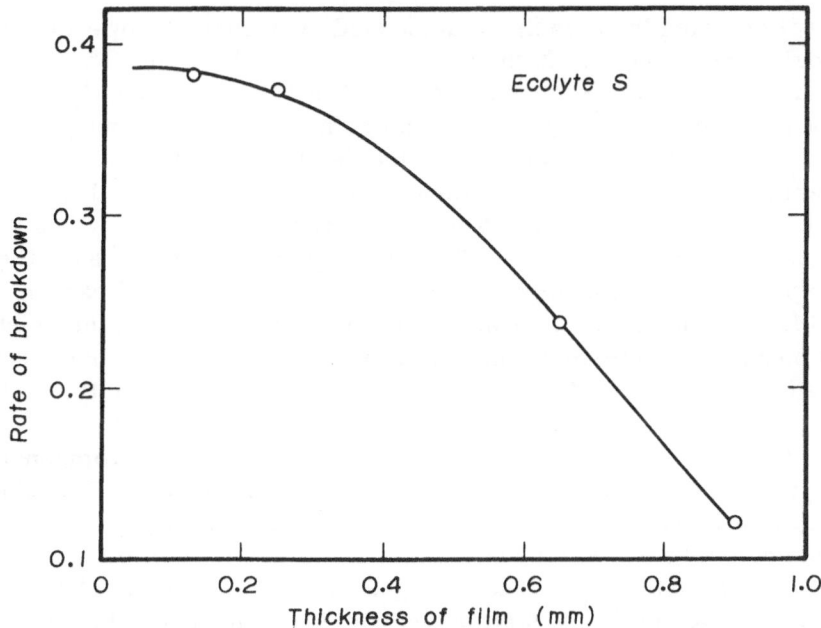

FIGURE 13.   Rate of degradation of ECOLYTE S as a function of film thickness.

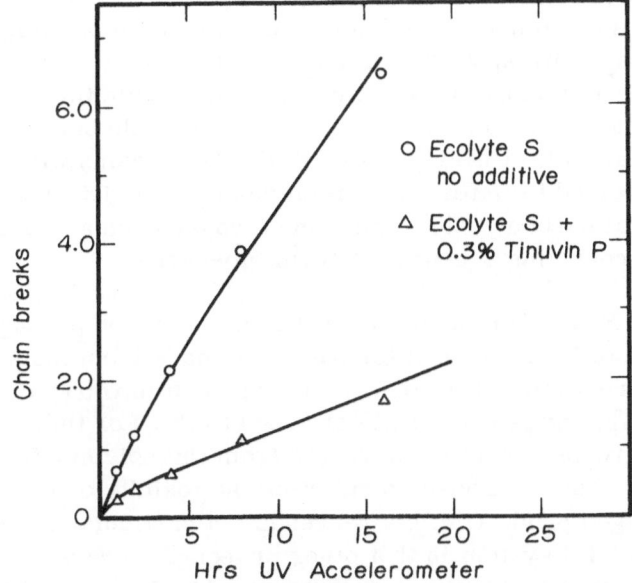

FIGURE 14.   Effect of UV stabilizer on degradation rate in ECOLYTE S.

controlled using the principles discussed previously. However, the
biological degradation which now ensues may take a considerably
longer time. The rate of this process depends on the chemical constitution of the original polymer, and as mentioned previously, on
the size of the particles and the molecular weight of the residual
material. It is possible to estimate the rate of biological degradation by various laboratory procedures, the most direct of which is
to mix the degraded product with a particular soil and measure the
biological oxygen consumption. By comparing the rate of carbon
dioxide evolution from the control soil and that containing the plastic,
it is possible to determine the rate at which the plastic is being converted to carbon dioxide and water by micro-organisms in the soil.

When photodegraded ECOLYTE P (modified poly(propylene))
is treated in this manner, the results are shown in Figure 15. One
can calculate on the basis that all of the poly(propylene) would be
biologically degraded in about a year under these conditions.
ECOLYTE S, on the other hand, degrades some ten times slower
than poly(propylene), even after photodegradation, and it is estimated that its lifetime until complete biological degradation would
be somewhere between five and ten years. However, the rate of
biological degradation is not crucial, and in fact it is desirable
not to have too rapid a degradation rate, otherwise one might expect either a brown spot or a green spot wherever the plastic
sample lay. For most purposes it seems desirable that a relatively
slow degradation should take place once the sample becomes part
of the natural soil. Ultimately the ECOLYTE resins will be completely converted to water and carbon dioxide, the latter returning
to the natural carbon cycle. This can be considered as a form of
"biological recycling", hence the term bio-cyclic.

Since ECOLYTE plastics will be used for food packaging,
the question of food approval for packages made from them is important. Essentially all of the ketone has been introduced into the
polymer chain and is chemically attached to it. For this reason,
the ketone groups cannot be extracted from the polymer film or
package and hence can have no effect on the toxicity or the taste
of the packaged product. This represents a particular advantage
of the ECOLYTE system in that other processes invariably make
use of additives which are merely dissolved and are not chemically

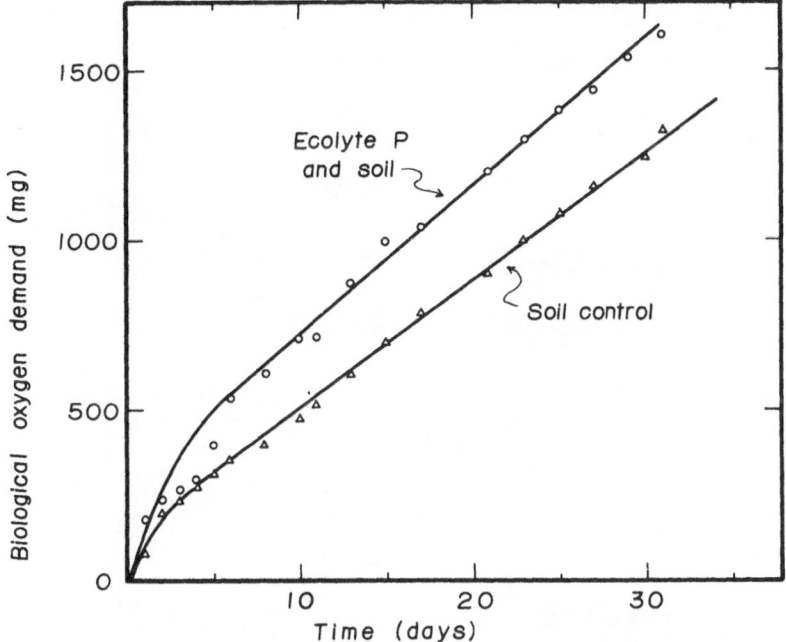

FIGURE 15. Biodegradation of photodegraded ECOLYTE P in soil tests.

bonded into the plastic and therefore will migrate from the plastic into any foodstuff packaged in it. Parenthetically, this point has been confirmed by the receipt of approval from the Canadian De-partment of National Health and Welfare, Food Advisory Bureau, for the use of ECOLYTE S in a variety of food packages. This represents the first photodegradable plastic to receive official sanction for packaging purposes. Although it may be possible to find food-approved additives, it is conceivable that these will effect the odor or flavor of the product packaged.

A practical example of the use of ECOLYTE S in foamed poly(styrene) drinking cups is shown in Figure 16. In this experi-ment ECOLYTE cups were attached by a small length of fishing line to anchors near the shore of a lake. After two weeks exposure to summer weather conditions the samples had completely broken up into small particles and disappeared. Control samples of

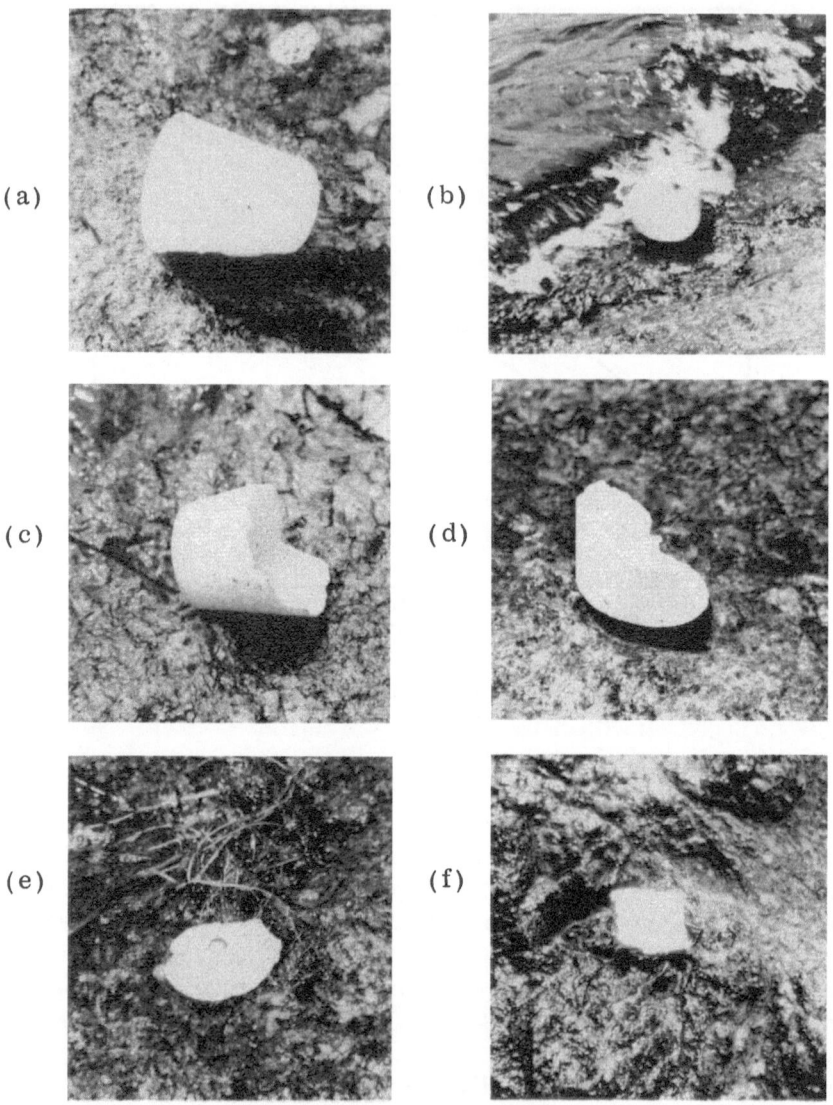

FIGURE 16. Exposure of ECOLYTE S foamed poly(styrene) cups on Canadian beach (Stoney Lake, Ontario), July 1972: (a) 1 day exposure; (b) 6 days; (c) 7 days; (d) 14 days; (e) 16 days; (f) 21 days. Note: cups were anchored on beach with thumbtack attached to a four foot length of nylon fishing line. Control cup of foamed poly(styrene) showed no degradation after three months exposure.

ordinary poly(styrene) were still unchanged at the beginning of winter and are expected to last at least several seasons under these conditions. Paper beverage cups were still undegraded under these conditions also. On the other hand, the cups stored indoors behind window glass or under normal room illumination have now remained more than six months without loss in properties.

In summary, the ECOLYTE process represents an inexpensive and reliable way of manufacturing plastics with a controlled lifetime. The process is simple since the synthesis of the ketone monomers involves inexpensive intermediates, and minimal additional costs are necessary to introduce them into the plastic materials. The process does not introduce any toxic or potentially toxic materials nor does the addition of the ketone group alter the flavor or odor of the plastic or the packaged products. After photodegradation the plastics degrade slowly by biological mechanisms, ultimately to CO_2 and water. The method is applicable to a wide variety of plastic materials of particular interest in the packaging industry.

REFERENCES

1.   G. H. Hartley and J. E. Guillet, Macromolecules, 1, 165 (1968).

2.   E. Dan, A. C. Somersall, and J. E. Guillet, Macromolecules (in press).

3.   F. J. Golemba and J. E. Guillet, Macromolecules, 5, 63 (1972).

4.   Y. Amerik and J. E. Guillet, Macromolecules, 4, 375 (1971).

5.   E. Dan and J. E. Guillet, Macromolecules (in press).

DELAYED ACTION PHOTO-ACTIVATOR FOR THE DEGRADATION OF PACKAGING POLYMERS

Gerald Scott

Department of Chemistry
The University of Aston in Birmingham
Gosta Green, Birmingham B4 7ET, England

SUMMARY

Fundamental chemical studies of the mechanisms of anti-oxidant action have led to the development of anti-oxidants which act by catalytically destroying hydroperoxides during polymer processing and subsequent exposure to ultraviolet light. They are, however, destroyed at the end of the photo-oxidation induction period and subsequently catalyze the photodegradation process. This type of system has considerable potential as a delayed action photo-initiator for the environmental destruction of plastics.

Until comparatively recently there was no interest in making polymers less stable in the environment. Indeed all the efforts of polymer technologists over many years have been directed toward improving the stability of the common polymers both during processing and subsequently during environmental exposure. Consequently extensive research has been in progress, mainly by industrial laboratories, to develop new stabilizing systems for polymers which would permit them to be used more extensively in the building and automotive industries.

In the course of these studies it has been found necessary to achieve a deeper understanding of the chemical mechanisms

27

involved in the oxidative degradation of polymers and particularly
of the way in which chemicals interfere with this process. As a
result, anti-oxidant and stabilizer technology has been elevated
from an empirical art to a sophisticated science. The basic
mechanisms involved in anti-oxidant action have been clearly
recognized and documented. [1, 2]

To the polymer scientist, therefore, the problem has been
not to make plastics less stable, but to improve their stability
during manufacture and under service conditions. In one sense
he has done his job too well. To the layman, plastics are an outstanding example of man-made materials which, due to their
resistance to biodegradation, have a deleterious effect on the
environment. The main materials which polymer scientists are
now replacing, for economic and technical reasons, are cellulose-
based. Due to its chemical structure cellulose biodegrades readily
after discard and is rapidly assimilated by the natural environment.
Plastics, on the other hand, are not readily wetted by water (which
is one of their advantages as a packaging material) and consequently
are not attacked by bacteria, [3] and the more common packaging
materials persist for a very long time in the environment. [4]

The photographs in Figure 1 illustrate the severity of the
problem of coastal pollution by plastics packaging. This formed
part of a collection of packaging litter picked up on a 50 yard
stretch of shoreline in north-west Scotland. The significance of
this site was that there was no habitation of any kind within a one
mile radius and the surrounding area was in any case very sparsely
populated. The sea-shore was quite inaccessible to the tourist
due to the steepness of the surrounding cliffs. Table I gives a
breakdown of the packaging waste identified. [4]

All the packaging had floated in from the sea and 20% of it
was not of British origin, but had originated in the European con-
tinent, the United States, and Canada. Much of it had lain undis-
turbed for many years. The low density poly(ethylene) (LDPE)
pack (Figure 1(a)) was of advanced age as evidenced by the high
carbonyl content of the polymer measured by ATR spectroscopy
and also based on the fact that the red closure had completely
faded on the side exposed to the sunlight. However, the plastic

(a)   (c)

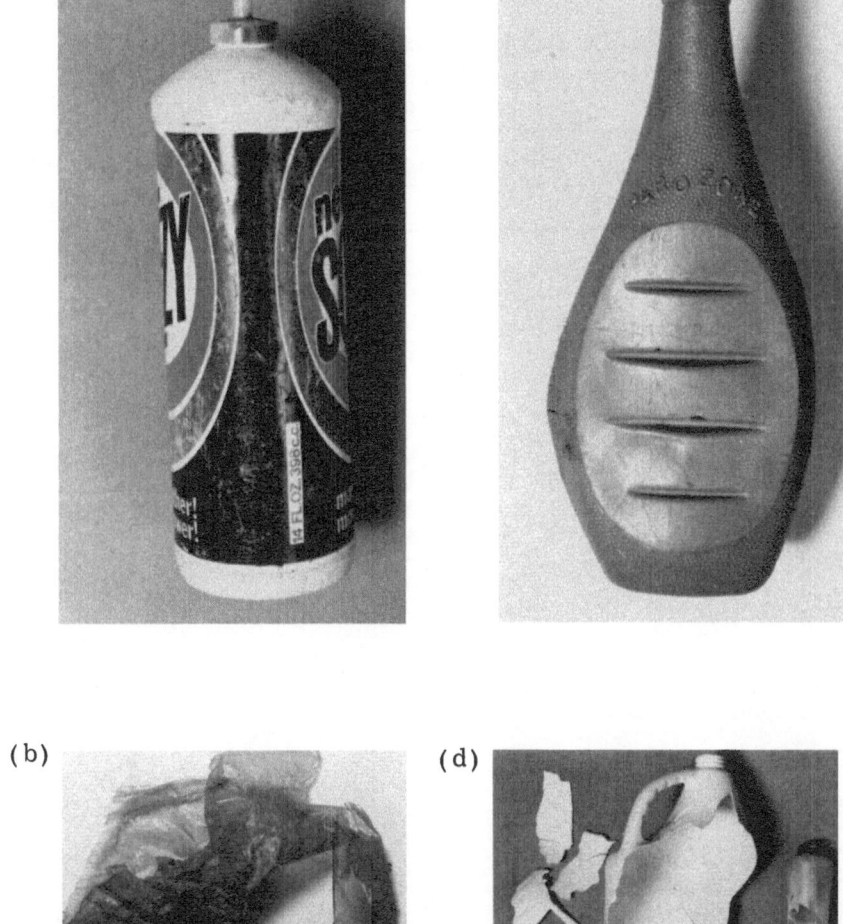

(b)   (d)

FIGURE 1. Examples of plastics packaging litter collected on a
50 yard stretch of shoreline in north-west Scotland.

TABLE I

PLASTICS PACKAGING IDENTIFIED AT
TWO SITES ON LOCH SCAVAIG

Package	Site A	Site B	Material of construction
Detergent containers	15	7	LDPE
Bleach and sanitary fluid containers	13	15	HDPE
Oil containers	3	4	HDPE
Cosmetic containers	13	3	LDPE & HDPE
Aerosol caps	1	1	ABS resin
Buckets and dishpans	7	2	LDPE
Food containers	—	1	ABS resin
Fertilizer sacks	3	2	LDPE
Thick gauge sheets	—	6	LDPE
Thin gauge sheets	2	—	LDPE
Ropes	—	6	Poly(propylene)
Fishing nets	—	2	Poly(propylene)
Miscellaneous unidentified	6	3	Mainly poly-(ethylene)

was as tough and flexible as when it was first made and would have
survived the weather for many years longer. It is probably true
to say that every detergent container of this kind ever made, which
has been deposited on the sea-shore or in the countryside, is still
there. The same applies to fertilizer sacks and bulk food containers. The farmer is a victim of circumstances in that his
fertilizers are delivered in packages which cannot be disposed of
after use. It is too expensive for the cleansing authorities to
collect them from him and they are normally discarded in the
fields where they accumulate. Heavily pigmented high density
poly(ethylene) (HDPE) bleach containers (Fig. 1(c)) were also
unchanged by extensive outdoor exposure although transparent
detergent containers made from the same polymer showed evidence
of embrittlement and breakdown over a long period of time (see
Fig. 1(d)). Moreover, white pigmented containers were less
weather resistant than colored packs (compare Fig. 1(d) with Fig.
1(c)). Figure 1(d) provides evidence that packaging plastics can,
under certain circumstances, break down in the environment over
a long period of time. Unfortunately this process is not fast
enough at the present rate of production of plastic litter.

Although the full extent of the problem has only recently
been recognized, it is clear that enormous tonnages of plastics
in the form of packaging are not collected into waste disposal
systems. One company in the United Kingdom has estimated that
this approaches 50%. [5] Much of this is made up of plastic sacks,
detergent containers and bleach containers. Foodstuffs packages
are generally much smaller than this in weight but may be just
as important from the point of view of nuisance value.

The 1971 UK figure for the use of LDPE in various disposable applications and the extrapolation to 1975 are given in Table
II together with the USA figures for 1971. [6] These figures do not,
of course, take into account any major new development such as
the introduction of plastic milk bottles on a large scale in the UK.
It has been estimated that if every household in the country were
to turn over to plastic non-returnable containers this would result
in a doubling of the amount of waste plastic packaging and a com-
mensurate increase in plastic litter. [7]

TABLE II

USE OF LDPE IN PACKAGING
IN THE UK AND USA (TONS)

Application	UK 1971	UK 1975	USA 1971
Disposable film and sheet	148,500	230,000	682,000
Disposable blow moulding	20,625	45,000	95,000
Disposable extrusion coating	9,500	15,000	197,000
	178,625	290,000	974,000

There can be no doubt therefore that the problem is a serious
one and is made all the more so by the fact that some of the newer
polymers in the packaging field are particularly resistant to environmental breakdown. In Sweden, almost no glass bottles are now
used for beer packaging. These have been replaced by the virtually indestructable poly(vinyl chloride) (PVC) pack. Similarly
in the USA, glass soft-drink bottles are being replaced by the new
environmentally-resistant poly(acrylonitrile) based materials
which have been shown in our own tests to be almost as UV resistant as PVC.

This situation threatens the future of plastics in packaging
and presents the polymer scientist with a powerful challenge. The
ideal packaging material is one which disappears in the environment at a similar rate to the more traditional organic materials
such as paper, cellophane, etc., which still retains the very good
barrier properties of plastics, and involves no extra cost to the
packaging manufacturer. A further requirement which is important in practice is that the polymer degradation process should not
be triggered off until the useful life of the package is finished. [8]
That is, it must have a safety period during which, even if it were
exposed to sunlight, it would not degrade.

The economic criterion ruled out as a general solution the possibility of synthesizing new polymers since one of the main attractions of plastics in packaging is their relative cheapness compared with other packaging materials, and these are cheap because of their scale of production. If it were necessary to devise a new range of plastics for each application, the scale of production of each one would be too small to allow them to compete effectively in cost with the major packaging polymers.

The use of an additive or series of additives which would induce UV instability into a range of packaging polymers seemed much more attractive. However, economic requirements introduced the further limitation that any such additive should not interfer with the productivity of the processing operation.[8] That is, the additive should be a photoactivator for the polymer during outdoor exposure, preferably after an induction period, but it should be inert during the processing operation.

Recent fundamental studies have shown that certain UV stabilizers for polymers function by removing hydroperoxides both during UV exposure and during the high temperature heat treatment of the polymer.[9] Hydroperoxides are key initiators for the process of oxidative degradation of polymers both during thermal-oxidative degradation and UV-initiated degradation. The basic chain reaction is the same in both cases.

$$2\,ROOH \longrightarrow ROO\cdot + RO\cdot + H_2O \left.\begin{array}{l}\\\\\\\\\\\\\end{array}\right\}$$

or $\quad ROOH \xrightarrow{\ h\nu\ } RO\cdot + \cdot OH$

$$\left.\begin{array}{l}RO\cdot\\ROO\cdot\\\cdot OH\end{array}\right\} + RH \longrightarrow \left\{\begin{array}{l}ROH\\ROOH + R\cdot\\H_2O\end{array}\right. \qquad\qquad \text{Initiation}$$

$$\left.\begin{array}{l}R\cdot + O_2 \longrightarrow ROO\cdot\\ROO\cdot + RH \longrightarrow ROOH + R\cdot\end{array}\right\} \begin{array}{l}\text{Chain}\\\text{reaction}\end{array}$$

$$2\,ROO\cdot \longrightarrow \text{Inert products} \qquad\qquad \text{Termination}$$

The main difference between the thermal and UV initiated
degradation lies in the rate of hydroperoxide breakdown. However, there is no doubt that powerful hydroperoxide decomposers
which remove hydroperoxides by a process which does not involve
free radical formation are both thermal and UV stabilizers.
Figure 2 shows the effect of a variety of metal dialkyldithiocarbamates on the UV stability of LDPE which has been subjected
to severe processing conditions.[9] It can be seen that at the end
of the milling operation, and after compression moulding the
polymer samples to film, all the dithiocarbamate-containing
films had oxidized to a lower degree than the control containing
only a conventional stabilizer as evidenced by the carbonyl content
of the film. They were therefore all powerful anti-oxidants for
the polymer, and this was confirmed by measuring the change in
melt-flow index of the polymer during the processing operation.
This was found to be unchanged even for the iron complex after
30 minutes milling, whereas poly(ethylene) containing a normal
organic soluble iron salt (e.g., ferric stearate or ferric acetylacetonate) was much more severely oxidized than the control under
the same conditions (see Fig. 3).[10] The effect of UV exposure
on the dithiocarbamate stabilized polymers was quite different.
Whereas the nickel and cobalt complexes stabilized the polymer
against the effects of UV light, the ferric and copper complexes
which are much less thermally stable were rapidly destroyed by
UV light. This could be followed by measuring the intensity of
the UV spectra of the metal complexes photometrically. The
nickel and cobalt complexes were found to have survived the
thermal oxidative treatment and the spectra did not disappear in
the polymer until after 22 hours of UV exposure during which
time the polymer was effectively UV stabilized. The UV spectra
of the iron and copper complexes, on the other hand, could not
be observed after processing although a slight color indicated
that both complexes were present in low concentration ($< 0.05\%$).
Even this color disappeared after four hours and the polymers
underwent rapid photo-oxidation, the iron oxidizing to embrittlement in about 120 hours, whereas the control was not brittle after
700 hours.

Fundamental studies of the mechanism of the anti-oxidant
action of the metal dithiocarbamates has shown[11] that they act

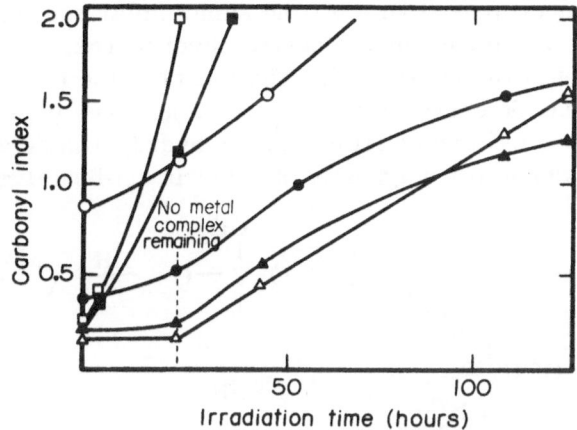

FIGURE 2. Effect of metal dithiocarbamates on carbonyl content
of poly(ethylene) during processing and accelerated UV exposure.
Concentration of metal complex = 2 x 10^{-3} moles/100 g.
☐ Fe(III) DEC; ■ Cu(II) DBC; ○ commercial polymer (con-
taining phenolic anti-oxidant only); ● Zn(II) DBC; ▲ Co(II) DEC;
△ Ni(II) DBC.

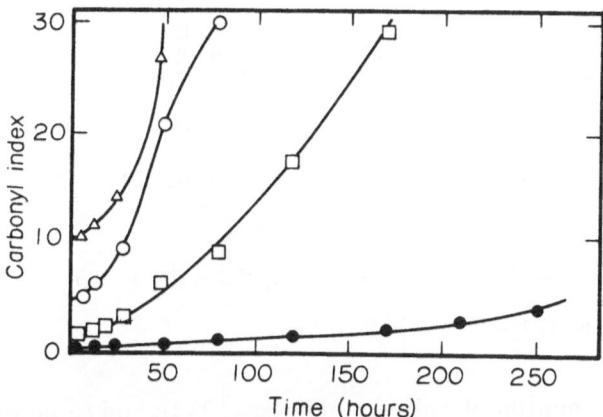

FIGURE 3. UV degradation of low density poly(ethylene) containing Fe(III) acetylacetonate. Concentration in moles/100 g:
△ 2 x 10^{-3} (brittle fracture, 80 hrs); ○ 1 x 10^{-3} (brittle fracture, 200 hrs); ☐ 8.5 x 10^{-5} (brittle fracture 250 hrs); ● no
additive (brittle fracture > 1000 hrs).

by generating a Lewis acid which acts as a catalyst for the destruction of hydroperoxides. At the same molar concentration,
all the metal dithiocarbamates destroy hydroperoxide in a first
order reaction at the same rate. The length of the induction
period to peroxide destruction does vary, however (see Fig. 4).
[12] The products formed from cumene hydroperoxide are phenol
and acetone which are diagnostic of a Lewis acid catalyzed process.

The Lewis acid in this case is believed to be SO_2 or possibly SO_3 which have been shown to be formed by the oxidation
of the dithiocarbamate complexes.

The formation of sulphur dioxide is believed to be the reason
for the very powerful anti-oxidant activity of the dithiocarbamate
complexes under processing conditions in the polymer, but unlike
some other peroxide decomposing anti-oxidants, the dithiocarbamate also reacts rapidly with hydroperoxides at ambient temperatures. This is almost certainly the main reason for the good UV
stabilizing activity of this class of anti-oxidants.[1] Not

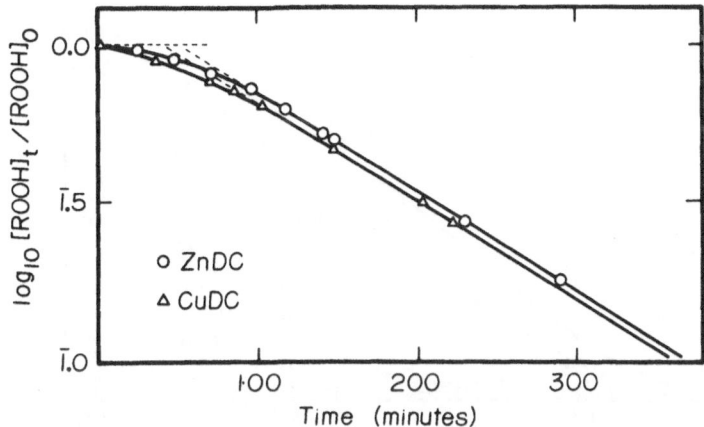

FIGURE 4. First order plot for the decomposition of cumene hydroperoxide in benzene (2×10^{-1} mol/l^{-1}) by metal diethyl-dithiocarbamates (2×10^{-4} mol/l^{-1}).

surprisingly then it is found that at higher concentrations the ferric dialkyldithiocarbamates are effective UV stabilizers. With decreasing concentration, the behavior of these metal complexes changes from that of a UV stabilizer through a delayed action UV activator (in which the rate of UV degradation during the induction period is lower than it is for the control sample), to a powerful UV activator (Fig. 5). This unusual behavior is characteristic of a large class of peroxide decomposing antioxidants based on sulphur and nitrogen ligands and is quite different from the energy transformers of which benzophenone and a variety of substituted benzophenones are typical. [13, 14] These are activated to the triplet (diradical) state by UV light and have the ability to hydrogen abstract from the polymer, followed by the expected radical reactions of the polymer radical so produced. In the presence of oxygen, this initiates the normal free radical chain process. As might be expected, the rate of oxidation initiated by the benzophenone triplet will be related to concentration in a different way. In fact, very much higher concentrations are required to achieve a rate of oxidation similar to that found with the metal complex activators (see Fig. 6). Moreover, instead of being auto-accelerators

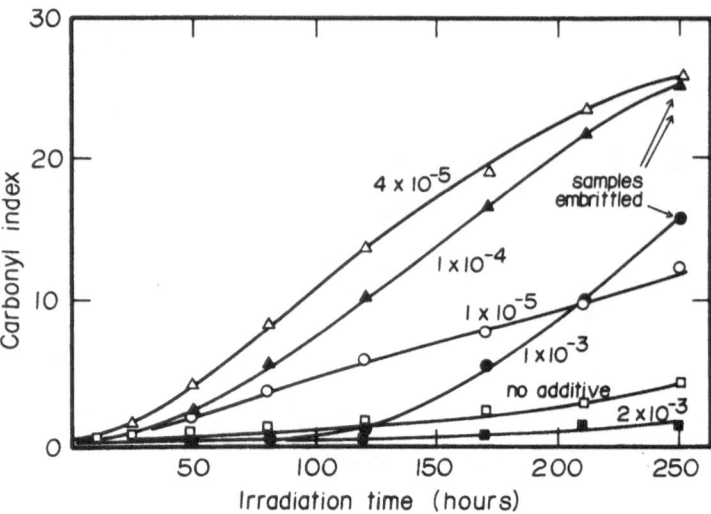

FIGURE 5. Effect of Fe(III) D9 DC on carbonyl formation in low density poly(ethylene). Numbers on curves are moles per 100 g of polymer.

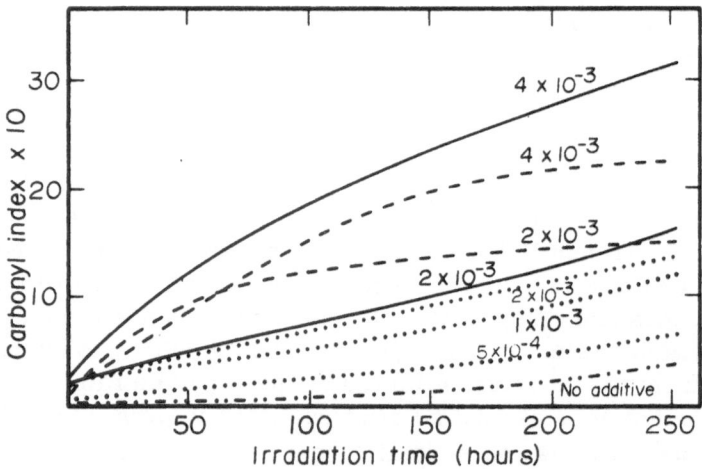

FIGURE 6. Effect of substituted benzophenones on carbonyl formation in low density poly(ethylene). Numbers on curves are moles per 100 g of polymer. —— 4, 4'-dichlorobenzophenone; — — — 4, 4'-dimethoxybenzophenone; · · · · 4-chloro-4'-methoxybenzophenone; — ·· — ·· — no additive.

of oxidation, these materials in the later stages lead to auto-retardation and under comparable conditions much longer times are required for embrittlement, even at high concentrations, than are possible with the metal complexes.

The <u>overall</u> effect of concentration of ferric dibutyldithio-carbamate on embrittlement time is shown in Figure 7. [9] The reason for the optimum is that at concentrations between 0.01 and 0.05%, the induction period is minimal (see Fig. 5). At concentrations around 0.1%, quite substantial induction periods are obtained in polymers processed under industrial processing conditions. This process is proving of considerable commercial significance[15] because as described earlier, the packaging user requires a certain minimum useful lifetime in his products and this can be arranged by adjusting the concentration of the photo-activator. Figure 8 compares the change of dynamic modulus with time of LDPE containing a photo-stabilizing system (cobalt dinonyldithiocarbamate) with LDPE containing a delayed action activator (iron dinonyldithiocarbamate Fe(II) D9DC). Whereas the dynamic modulus of the polymer remains unchanged and less than the control during prolonged UV exposure, the delayed-action system shows an initial delay time before changing auto-catalytically. [14] The delayed action UV activator system is not limited to sulphur ligands. A ferric complex not containing sul-phur acts in a parallel way to the dithiocarbamate complexes. This effect is shown in Figure 9. Again the nickel complexes are well known UV stabilizers[16] whereas the ferric complex is a delayed action activator at low concentrations. [14] Two component

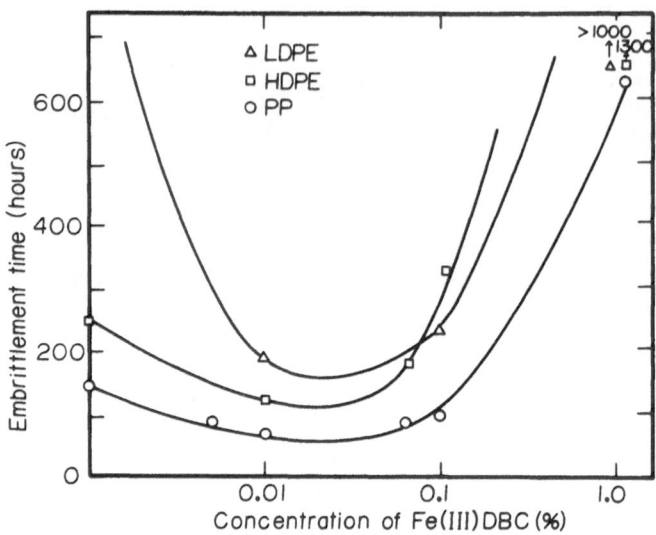

FIGURE 7. UV stability of poly(propylene), low density poly-
(ethylene) and high density poly(ethylene) films (240 μm) containing Fe(III) DBC.

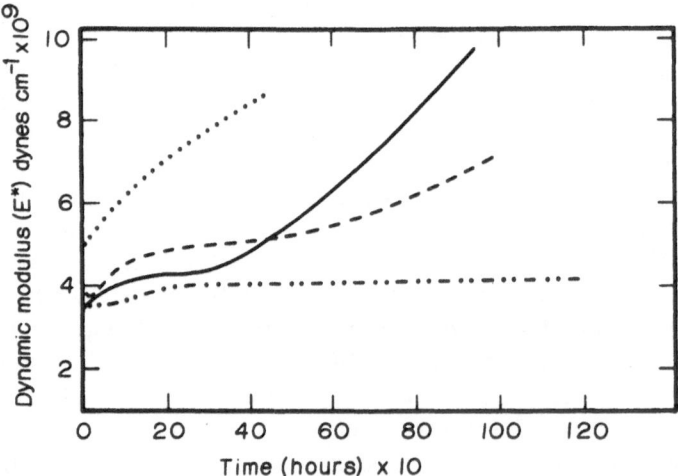

FIGURE 8. Effect of dithiocarbamates on the dynamic mechanical
properties of low density poly(ethylene) on UV exposures. · · · ·
ionic Fe^{3+}; ——— Fe(II) D9 DC; – – – no additives; –··–··–
Co(II) D9 DC.

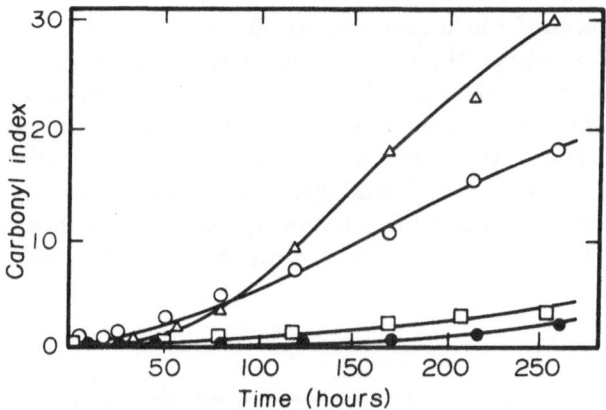

FIGURE 9. UV degradation of low density poly(ethylene) containing

$$\begin{bmatrix} CH_3-\langle\!\langle O \rangle\!\rangle-O \\ \qquad\quad C=N \\ \quad CH_3 \quad OH \end{bmatrix}_3 Fe(III)$$

Concentrations in moles/100 g: △ 9×10^{-5} (brittle fracture 260 hrs); ○ 2×10^{-5} (brittle fracture 300 hrs); □ no additive (brittle fracture > 1000 hrs); ● 1×10^{-3} (brittle fracture > 1000 hrs).

delayed action activators consisting of a transition metal ion with
a complexing agent to restrain the metal ion during processing
are proving to be very useful because of the wide range of delay
times and photodegradation rates that can be achieved. Figure
10 shows three such systems based on ferric stearate at the same
molar concentration and molar ratios of complexing agent to
metal ion. [14]

Recent work has shown that when the plastic breaks down
to a friable powder, the molecular weight is reduced from approximately 80, 000 to about 40, 000. At this point, the powder is water-
wettable but shows little sign of biodegradation. However, the
molecular weight is reduced still further by oxidative processes
(which occur even in the absence of light due to the presence of
the transition metal ion) and biodegradation will ultimately occur
when the molecular weight falls below 5, 000. [17] At this point
the material becomes effectively assimilated into the environment.

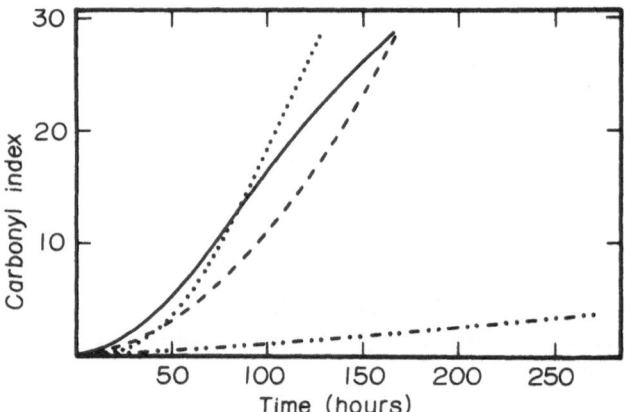

FIGURE 10. Two component UV activator systems for low density poly(ethylene). Initial concentration of each additive, 0.1%.
· · · · System A + FeSt,(brittle fracture 250 hrs); —— System B
+ FeSt (brittle fracture 290 hrs); − − − System C + FeSt (brittle
fracture 300 hrs); −··−··− no additive (brittle fracture >1000 hrs).

REFERENCES

1. G. Scott, Atmospheric Oxidation and Antioxidants, Elsevier, New York, 1965.

2. W. L. Hawkins, Ed., Polymer Stabilization, Wiley-Interscience, New York, 1972.

3. H. O. W. Eggins, J. Mills, A. Holt, and G. Scott, in Microbial Aspects of Pollution, Eds. G. Sykes and F. A. Skinner, Academic Press, London, New York, 1971.

4. G. Scott, Internatl. J. Environ. Studies, 3, 35 (1972).

5. Imperial Chemical Industries Publications for Schools, Plastics (Part 3), 1972, p. 13.

6. International Technical Surveys, Feasibility Study of Controlled Degradable Plastics, 1972.

7. A. E. Higginson, Plastics and Polymers Conference Supplement (Plastics Institute), No. 4, Sept. 1971.

8. G. Scott, Plastics, Rubbers, Textiles, 1, 361 (1970).

9. G. Scott, Symposium on Macromolecules, Helsinki, 1972, Special Publication (in press).

10. D. C. Mellor, A. Moir, and G. Scott, Eur. Polym. J. (in press).

11. J. D. Holdsworth, G. Scott, and D. Williams, J. Chem. Soc., 4692 (1964).

12. A. Rawlinson and G. Scott, unpublished work.

13. G. Scott, Plastics and Polymers Conference Supplement (Plastics Institute), No. 4, (Sept. 1971).

14. M. U. Amin and G. Scott, unpublished work.

15.   G. Scott, Belgian Patent 770, 202.

16.   P. J. Briggs and J. McKellar, J. Appl. Polym. Sci., $\underline{12}$, 1825 (1968).

17.   H. O. W. Eggins, J. Mills, L. M. Gan, and G. Scott, unpublished work.

THE PHOTOACTIVATED DEGRADATION OF POLYOLEFINS

Bernard Baum and Roy A. White

DeBell & Richardson, Inc.
Hazardville Station
Enfield, Connecticut, 06082

ABSTRACT

Degradation of plastics under weathering is an oxidation catalyzed by the high energy of ultraviolet, and complicated by the erosion of wind, dust, and rain. Ultraviolet absorbers, and more recently zinc oxide, have long been used to protect plastics against UV degradation. Now, selected systems of photoactivators have been found that accelerate the degradation of plastics under UV. The effect of anti-oxidants and pigments on the rate of decomposition of photoactivated poly(ethylene) and poly(propylene) is shown. These prodegradant systems should be useful in mulch film and in solving the problem of plastics litter.

INTRODUCTION

A survey conducted by the Highway Research Board of the National Research Council[1] indicated that approximately one cubic yard of litter was accumulated per month, on the average, for each mile of interstate and primary highway in the 29 states participating in the survey. It was estimated that 59% of all items in this litter was paper, 16% was metal, 6% was plastic, 6% was glass and 13% was miscellaneous. The paper eventually decomposes, but the plastic, especially as film, is a highly visible and seemingly permanent component of this waste.

45

Biodegradation appears doubtful as a primary economic route for decomposing polyolefins, which make up the bulk of the packaging waste. This leaves only oxidative, ultraviolet degradation. It is ironic: for years chemists have worked to stabilize plastics against ultraviolet decomposition. Now they are being asked to do the opposite.

Our goal was to find inexpensive, FDA approvable additives which would accelerate the UV degradation of polyolefins in sunlight but not cause degradation indoors. The erosive forces of wind and rain would complete the process of breaking up the embrittled polymer into small particles.

The sun's spectrum at the earth's surface extends down to approximately 290 mµ. Ultraviolet at 300 mµ has 95 kcal of energy, enough to break the carbon - carbon bond. The wavelength of maximum sensitivity of poly(ethylene) lies at 300 mµ, that of poly-(propylene) at 310 mµ.

Although a pure paraffin would not be expected to absorb UV light, polyolefins contain small amounts of peroxide and carbonyl-containing impurities formed by oxidation during manufacture. Ultraviolet energy is absorbed, oxidation occurs, and the polyolefin molecule is dissociated.

The mechanism which follows[2] has been found to apply to low molecular weight hydrocarbons and by inference to poly(ethylene).

Initiation

$$\text{formation of } R \cdot \tag{1}$$

Propagation

$$R \cdot + O_2 \longrightarrow RO_2 \cdot \tag{2}$$

$$RO_2 \cdot + RH \longrightarrow RO_2H + R \cdot \tag{3}$$

Termination

$$R \cdot + R \cdot \longrightarrow R{-}R \tag{4}$$

$$R \cdot + RO_2 \cdot \longrightarrow RO_2R \tag{5}$$

$$2\,RO_2 \cdot \longrightarrow RO_2R + O_2 \tag{6}$$

The first reaction, initiation, could occur by the breakdown of
the hydroperoxide, RO_2H. This hydroperoxide can form either
by direct air oxidation of an activated weak spot in the resin[3]
or through propagation, as in Eq. (3). The latter effect, however, assumes more importance as auto-oxidation proceeds.
Equations (4) and (5) are not significant at the oxygen pressures
used in this work.[4] As seen by consideration of the foregoing
mechanism, oxidation is influenced by factors that catalyze peroxide decomposition. Such factors include ultraviolet radiation,
and certain metals and metallic ions.

The abstraction of H· by RO_2· yields the initiating radical,
R·, which is the basis of autocatalysis in all hydrocarbon oxidation. Accordingly, the relative oxidizability of any hydrocarbon
depends on the strength of the carbon-hydrogen bond. It is well
substantiated in the literature[5] that a tertiary hydrocarbon is
more susceptible to oxidation than a secondary hydrogen, which
in turn is more susceptible than a primary hydrogen. Also, adjacent double bonds[6] and carbonyl groups[7] tend to weaken the
strength of the C—H bond, probably primarily as a result of the
increased resonance stabilization of the R· radical formed.

Possible mechanisms for the decomposition of poly(propylene)
hydroperoxide are:[10]

All mechanisms of polyolefin oxidation lead to oxygenated products. Although decomposition of the hydroperoxide formed usually leads to chain splitting, recombination of free-radical products can lead to cross-linking in poly(ethylene). Any additive system developed must accelerate the chain splitting while retarding cross-linking.

EXPERIMENTAL

Some of the many parameters for determining the degree of weathering degradation include measurement of melt index, carbonyl, peroxide, carboxyl, intrinsic viscosity, changes in color, surface crazing or cracking, tensile strength, elongation, brittle temperature, angle of break in flexure, breaking when bent around a mandrel, impact strength, and electrical properties. The first change that occurs during hydrocarbon oxidation is the formation of hydroperoxide and peroxide, followed closely by increase in carbonyl content. Loss of physical properties first appears as a decrease in tensile elongation. In this study we have used carbonyl as measured by infrared, and tensile strength and elongation as parameters of degradation.

We use a multi-additive system, each component of which performs a different function. These additives are blended into the polymer on a two-roll mill, sheets are compression molded and exposed both under an RS-4 sunlamp and out of doors at Hazardville, Connecticut. The technique used for preparation of samples was essentially the same for all polyolefins. Additives were mixed into polymer by masticating for 10 minutes on a 6" x 12" differential-speed, two-roll mill. The resulting milled sheets were molded into flat test sheets 6" x 6" x 0.005" by preheating for one to two minutes, compressing, and holding under 1100 P.S.I. pressure for one minute, and cooling for one minute under full pressure before removing the sheet from the mold.

For outdoor weathering, samples measuring 1" x 2" x 0.005" were exposed (at Hazardville, Connecticut) in a fine wire cage at a 45° angle facing south.

Tensile strength and percent ultimate elongation were measured on micro-tensile specimens according to ASTM D-1708.

Carbonyl measurements were made at 5.84 μ with a Perkin-Elmer Model 21 (scanning speed, 1/2 micron per minute) using a compression-molded sheet in each beam of a double-beam monochromator. The reference beam has an unirradiated sample. Absorbance was determined as follows:

$$\text{Absorbance/mil} = \frac{\log \frac{I_0}{I}}{\text{thickness}}$$

where

I_0 = incident radiation, percent transmission

I = transmitted radiation, percent

Materials used were low-density poly(ethylene) (Union Carbide DYNH); poly(propylene) (Hercules Profax 6523); anti-oxidant (0.1% butyl zimate (zinc dibutyldithiocarbamate) Vanderbilt); titanium dioxide (Anatase) (A-Lo, National Lead, used at 2 phr[*] level); violet pigment (Blythe Mineral Violet (Violet 16), United Mineral & Chemical Corporation, used at 1 phr level); green pigment (Blythe Blue-Green (Pigment Blue 36), United Mineral & Chemical Corporation, used at 1 phr level)

RESULTS AND DISCUSSION

Poly(ethylene)

A number of additive systems have been examined for their efficiency as photoactivators in both poly(ethylene) and poly-(propylene). The accelerating effect of one such system is shown in low-density poly(ethylene) (LDPE) (Fig. 1) where percent elongation is plotted against outdoor exposure time. After one month of outdoor exposure, the sample containing photoactivator System 5 had lost 98% of its elongation and was brittle. The control still retained 91% of its original elongation.

---

* phr, parts additive per 100 parts resin.

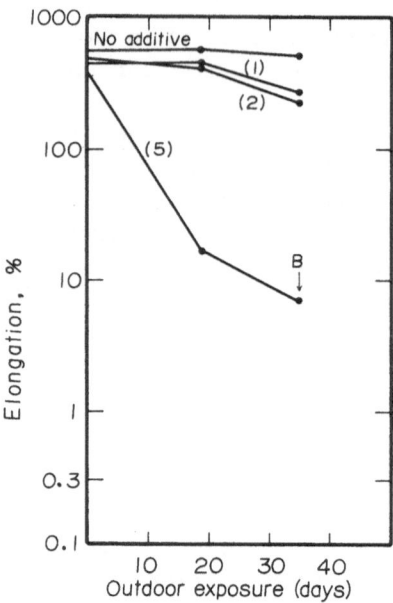

Figure 1. Effect of pigments in photoactivated poly(ethylene) on loss of elongation during outdoor exposure. (1) TiO$_2$ plus photoactivator no. 5; (2) violet pigment plus photoactivator no. 5; (5) photoactivated; B, breaks readily.

Two pigments, titanium dioxide and Blythe Mineral Violet, were used at 1 phr with Photoactivator 5. Both retarded the rate of decomposition.

During outdoor exposure (Fig. 1) there was a steady and continuous loss of elongation. Loss of tensile strength (during RS-4 sunlamp exposure, Fig. 2) is also a continuous process. Again, both pigments act as retarders.

Photoinitiated poly(ethylene) loses tensile strength and elongation on exposure to ultraviolet (of the sun) in the presence of air because of loss in molecular weight from chain splitting. This process is an oxidation accelerated by UV. In previous work by Baum, [8] hydroperoxide was detected as the first product of UV or thermal oxidation, followed closely by peroxide. As the hydroperoxide and peroxide decomposed, a complex mixture of other products (e.g., acids, ketones, alde-

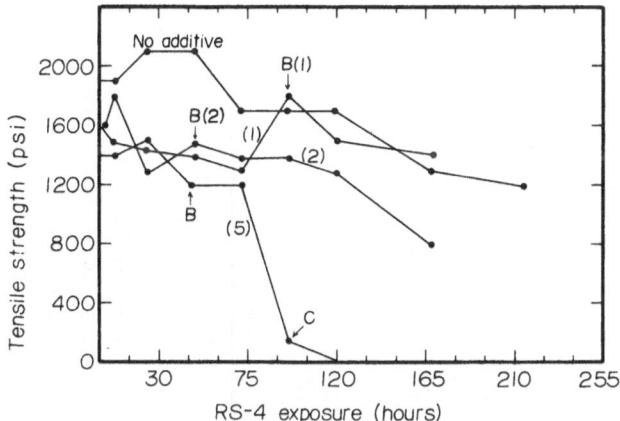

Figure 2. Loss of tensile strength during RS-4 sunlamp exposure: effect of pigments in photoactivated poly(ethylene). (1) TiO_2 plus photoactivator no. 5; (2) violet pigment plus photoactivator no. 5; (5) photoactivated; B, breaks readily; C, crumbles readily.

hydes, esters, and ethers) formed and could be detected quantitatively by infrared measurement of carbonyl at approximately 5.8 microns. Simultaneously there was a loss in molecular weight as indicated by increase in melt index.[9]

In Figure 3 carbonyl formation during outdoor exposure of poly(ethylenes) containing three different photoinitiator systems is compared to a control. Note the wide difference in efficiency among the three systems.

The loss of elongation and tensile strength of the compounds in Figures 1 and 2 was found to correspond to the increase in carbonyl. Carbonyl formation similar to that shown in Figure 3 is found when pigmented samples are exposed under the RS-4 sunlamp (Fig. 4). The rate of carbonyl formation is much greater when poly(ethylene) contains a photoinitiator (No. 5). Comparison of Figures 1 and 2 with Figure 4 reveals that the rate of carbonyl formation is only a valid parameter for loss of physical properties in virgin resin.

In Figures 1 and 2, loss of elongation and tensile strength of photoactivated LDPE containing pigments and exposed outdoors is much slower than it is with the photoactivator alone, and is similar

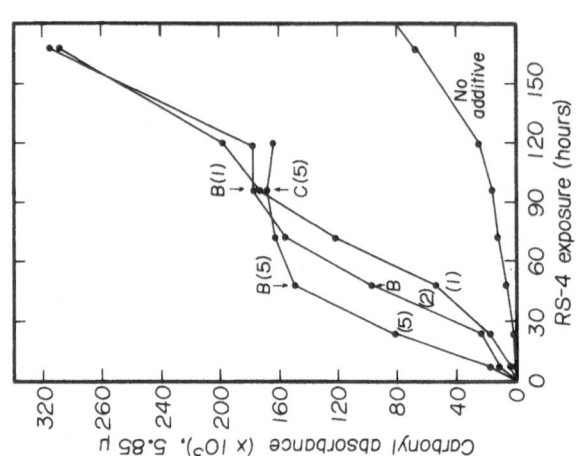

Figure 4. Pigments in photoactivated poly(ethylene): change in carbonyl absorbance during RS–4 sunlamp exposure. (1) TiO_2 plus photoactivator no. 5; (2) violet pigment plus photoactivator no. 5; (5) photoactivated; B, breaks readily; C, crushes readily.

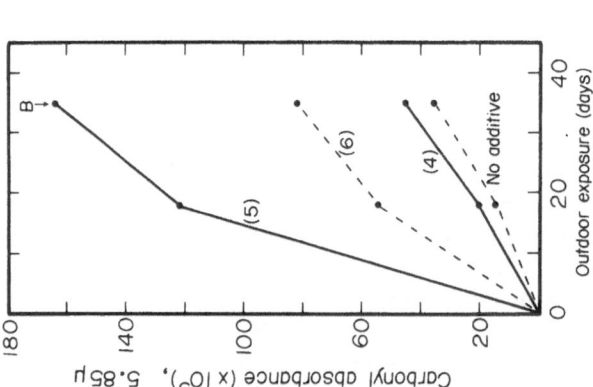

Figure 3. Change in carbonyl absorbance during outdoor exposure of photoactivated poly(ethylene). (4) photoactivated; (5) photoactivated; (6) photoactivated; B, breaks readily. Note: in all of this work, carbonyl was measured as absorbance per mil.

the control; whereas the rate of carbonyl formation under the RS-4
sunlamp of photoactivated LDPE with pigments is much faster
than in the control.

Under the RS-4 sunlamp, both No. 5 and No. 5 containing
titanium dioxide became brittle (B) in the same time span (47-
48 hours) (Figs. 2 and 4); on exposure out of doors, however
(Fig. 1), only No. 5 without pigments had embrittled, indicating
that accelerated exposure does not always correlate with outdoor
weathering.

Rate of carbonyl formation does not always appear to be
useful as a screening tool. Under the RS-4 sunlamp, No. 5
and No. 5 with either pigment reach the same carbonyl content
at about 96 hours; however, at 96 hours only No. 5 had crumbled
(C). At 168 hours and much higher carbonyl, neither of the No.
5's containing pigment had crumbled.

Figure 5 shows a plot of tensile strength vs. carbonyl for several
photoactivated samples exposed under the RS-4 sunlamp to determine if a critical carbonyl content exists. At the 160-180 absorbance there is a catastrophic failure for two of the systems (as
indicated by the C (crumbling point), but not for a third system
(No. 3). There is obviously a different mechanism of degradation
for No. 3. During oxidation, chain scission and cross-linking
occur simultaneously. Sample No. 3 may have had a higher
degree of cross-linking and thus held together beyond the crumbling point, indicated by the carbonyl content. Alternately, chain
scission may not necessarily correlate with rate of carbonyl
formation.

The initial rise in strength shown in Figures 2 and 5 may be
due to cross-linking.

The question always arises as to whether accelerated UV
aging can be used to predict the results of outdoor weathering.
During the previous discussion this did not necessarily appear to
be true in the presence of the titanium dioxide and violet pigments.
However, where only virgin resin or resin plus photoactivator

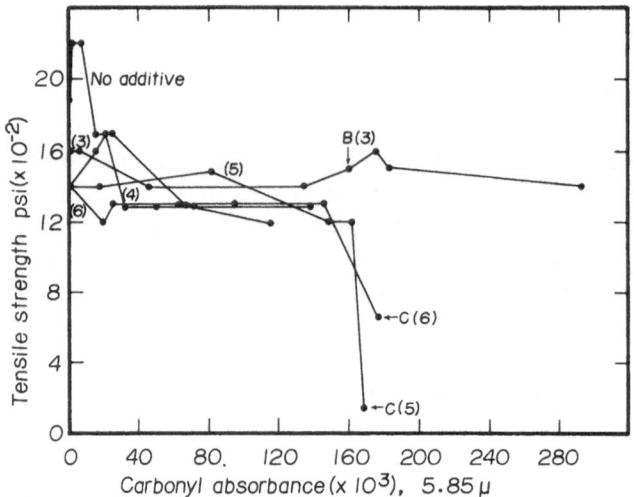

Figure 5. RS-4 sunlamp exposure: tensile strength vs. carbonyl absorbance in photoactivated poly(ethylene). (3) photoactivated; (4) photoactivated; (5) photoactivated; (6) photoactivated; B, breaks readily; C, crushes readily.

is concerned, there is a reasonable correlation between rate of loss of elongation under the RS-4 sunlamp and during outdoor (summer) exposure (Fig. 6).

Poly(propylene)

Effective photoactivator systems have also been found for poly(propylene) (Fig. 7). However, unlike poly(ethylene), only one of the pigments retards the rate of oxidation; the other, titanium dioxide, accelerates loss of tensile strength.

Rate of carbonyl formation (Fig. 8) in poly(propylene) correlates with rate of loss of tensile strength (Fig. 7). The green pigment slows the rate of carbonyl formation, while titanium dioxide accelerates it up to a point.

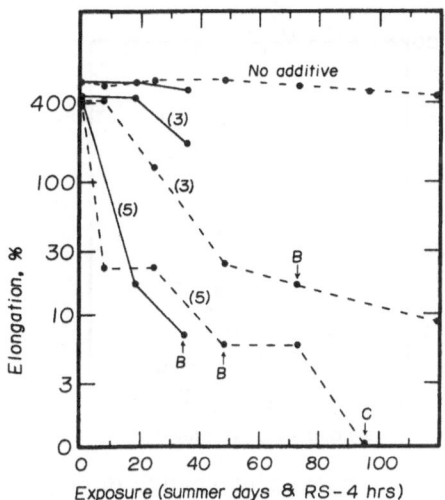

Figure 6. Loss of elongation during outdoor and RS-4 sunlamp exposure of photoactivated poly(ethylene). (3) photoactivated; (5) photoactivated; B, breaks readily; C, crushes readily; ━ ━ ━ RS-4; ━━━━━ outdoor.

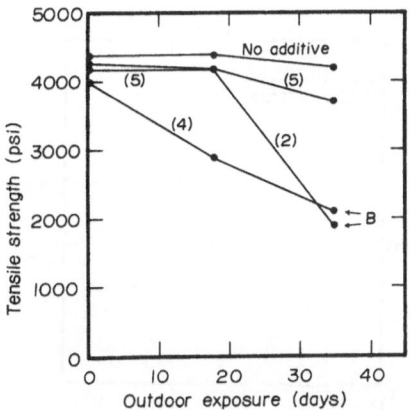

Figure 7. Effect of pigments on tensile strength in photoactivated poly(propylene) during outdoor exposure. (2) photoactivated; (4) TiO_2 plus photoactivator no. 2; (5) green pigment plus photoactivator no. 2; B, breaks readily.

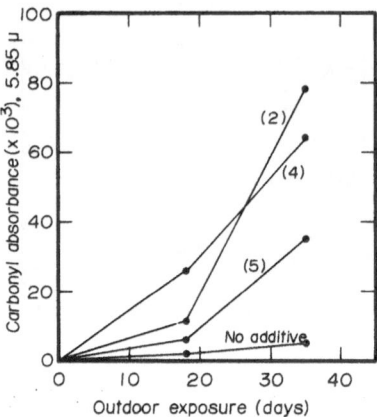

Figure 8. Change in carbonyl absorbance during outdoor exposure of photoactivated poly(propylene) containing pigments. (2) photo-activated; (4) TiO$_2$ plus photoactivator no. 2; (5) green pigment plus photoactivator no. 2.

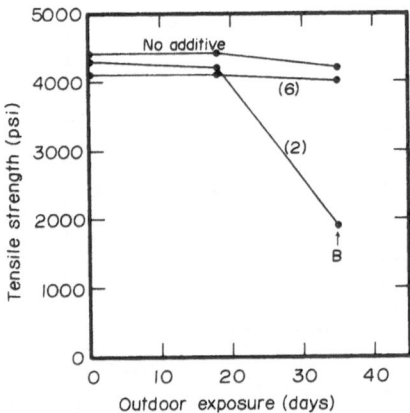

Figure 9. Tensile strength of photoactivated poly(propylene) con-taining antioxidant during outdoor exposure. (2) photoactivated; (6) antioxidant plus photoactivator no. 2; B, breaks readily.

An anti-oxidant (0.1% dibutyl dithiocarbamate) added to photoactivated poly(propylene) has the expected effect, inhibition of both rate of tensile loss (Fig. 9) and rate of carbonyl formation (Fig. 10). We have experienced similar results with phenolic, sulfur, and phosphite anti-oxidants, indicating that only minimal anti-oxidant content should be used.

Photographs (Figs. 11, 12 and 13) were taken of poly-(propylene) after the following exposures: Figure 11, photo-activated poly(propylene) after being exposed outdoors (at Hazard-ville, Connecticut) during the two summer months; Figure 12, photoactivated poly(propylene) after hanging indoors in the labora-tory under fluorescent lights near a window for two and one-half years; and Figure 13, a control sample (no photoactivator) after outdoor exposure (including summer) for eight months.

These photographs indicate the rapid rate of decomposition of poly(propylene) containing a photoactivator system. They also show that poly(propylene) containing the photoactivator system does not degrade when suspended under bright fluorescent lights indoors.

CONCLUSIONS

Inexpensive additive systems that accelerate the ultraviolet decomposition of low- and high-density poly(ethylene) and poly-(propylene) have been developed. Since this occurs by a free-radical mechanism, anti-oxidants, as expected, retard the reaction.

Both of the pigments examined retarded the decomposition of photoactivated poly(ethylene). In photoactivated poly(propylene) a green pigment also acted as a retarder, but Anatase titanium dioxide accelerated the decomposition. Retardation can be over-come either by keeping the anti-oxidant and pigment concentration at a minimum or by careful selection of pigments.

For a given formulation, increase in carbonyl content paral-lels the loss of tensile strength or elongation. The rate of carbonyl formation is a good screening test for virgin polymer but is, at best, only an approximation when pigments are present.

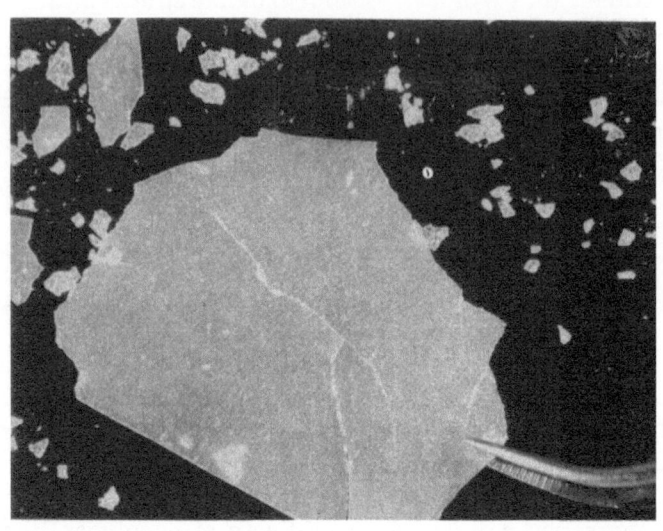

Figure 11. Poly(propylene plus additive, exposed outdoors for two summer months.

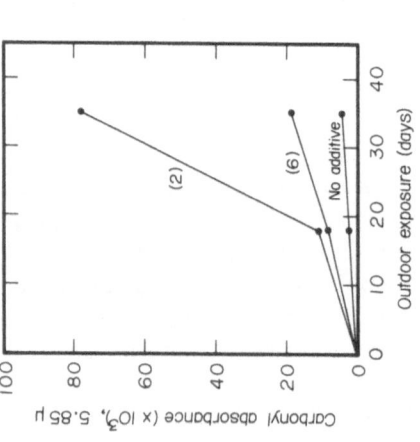

Figure 10. Change in carbonyl absorbance during outdoor exposure in photoactivated poly(propylene). (2) photoactivated; (6) anti-oxidant plus photoactivator no. 2.

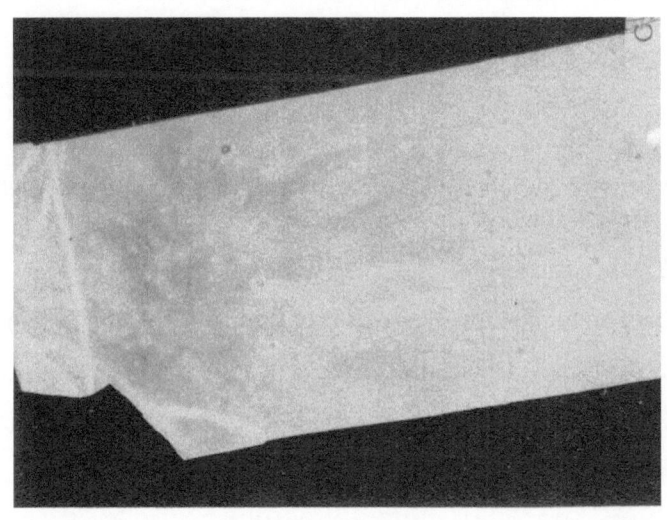

Figure 13. Poly(propylene) control after outdoor expsoure for eight months.

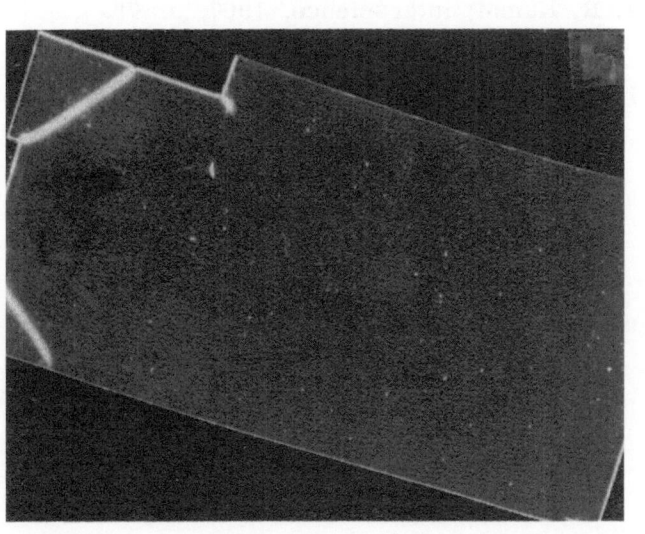

Figure 12. Poly(propylene) plus additive after exposure to fluorescent lighting in-doors for two and one-half years.

RS-4 sunlamp aging shows reasonable but not consistent correlation with outdoor weathering. Sunlamp data must be confirmed by outdoor weathering.

REFERENCES

1. "National Study on Roadside Litter", National Research Council, Highway Research Board, September 1969.

2. B. Baum, J. Polym. Sci., 2, 281 (1959).

3. L. Bateman, Trans. Inst. Rubber Ind., 26, 246 (1950).

4. J. L. Bolland, Proc. Roy. Soc. (London), 186, 230 (1946).

5. C. F. Cullis, et al., Disc. Farad. Soc., No. 2, 111 (1947).

6. R. B. Mesrobian and A. V. Tobolsky, J. Polym. Sci., 2, 463 (1947).

7. C. N. Hinshelwood, Disc. Farad. Soc., No. 10, 266 (1951).

8. B. Baum, unpublished work.

9. B. Baum, Plastics Technology, April 1961, p.29.

10. F. H. McTigue and M. Blumberg, Appl. Polym. Symposia, Ed. M. R. Kamal, Interscience, 1967, p.176.

THE BIODEGRADABILITY OF SYNTHETIC POLYMERS

J. E. Potts, R. A. Clendinning,
W. B. Ackart, and W. D. Niegisch

Union Carbide Corporation
River Road
Bound Brook, New Jersey, 08805

Of the large number of plastics produced commercially in this country, three account for about 90% of the plastic found in municipal waste.[1] These three are poly(ethylene) at 38%, poly-(vinyl chloride) at 31% and poly(styrene) at 21%. Because these three plastics make up the bulk of packaging plastics, it was felt that they should be the focus of attention of research and development pertaining to the disposability or recycling of plastic waste.

The literature in the field of biodegradability of synthetic polymers is almost entirely concerned with the problem of pre-venting or retarding attack on plastics by micro-organisms and with the susceptibility of plasticizers, etc. to attack.[2-7]

We have chosen to define biodegradable materials as those which, because of their chemical structure, are susceptible to being assimilated by micro-organisms such as fungi and bacteria. Some non-biodegradable plastics are erroneously believed to be biodegradable because they often contain biodegradable additives, which will support the growth of micro-organisms without causing the plastic itself to become assimilated. The term "biodegradable" is often used indiscriminately to refer to various types of environ-mental degradation, including photodegradation. Because a poly-meric material is degraded by sunlight and oxygen does not mean that the material will also be assimilated by micro-organisms.

61

The term "biodegradable" should be reserved for that type of
degradability which is brought about by living organisms, usually
micro-organisms.

MICROBIOLOGICAL PROCEDURES

Samples were tested for degradation by fungi, using ASTM
Method D-1924-63. This procedure requires the placement of
test specimens in or on a solid agar growth medium that is defi-
cient only in carbon. Any growth which may occur is dependent
on the utilization of a component of the specimen as a carbon
source by the test organism.

High molecular weight polymer samples were examined as
strips cut from compression-molded plaques or as finely ground
powders dispersed in the agar. Semisolid waxes, greases, and
organic liquids were deposited on biologically inert fiberglass
cloth which was then placed on the agar.

The test fungi consisted of a mixture of Aspergillus niger,
Aspergillus flavus, Chaetomium globosum, and Pencillium funicu-
losum. After an exposure time of three weeks, samples were
examined and assigned growth ratings as follows: 0, no growth;
1, traces (less than 10% covered); 2, light growth (10 - 30%
covered); 3, medium growth (30 - 60% covered); and 4, heavy
growth (60 - 100% covered). Strips of filter paper served as controls in each experiment to ensure that an active fungal mixture
was used.

In addition to the agar plate methods, some plastic samples
were buried in a mixture of equal parts by volume of garden soil,
builders sand, and peat moss, which was placed in flower pots and
kept moist with water.

BIODEGRADABILITY OF COMMERCIAL PLASTICS

We have investigated the biodegradability of a large number
of commercial plastics with an emphasis on those used in packaging.

Using the ASTM test nearly all are resistant to attack as shown in Table I. Several samples which did show susceptibility to attack showed no growth after the sample was extracted with solvent. This behavior indicates that the original sample has a biodegradable additive and that the polymer molecule itself is not attacked. All of the large volume packaging plastics based on poly(ethylene), poly-(styrene) and poly(vinyl chloride) did not show any susceptibility to attack.

The poly(urethane) in Table I which gave a growth rating of 4 is based on a poly(ester) diol. This result is in agreement with the work of Darby and Kaplan[6] who found that poly(urethanes) based on poly(ester) diols were more susceptible to attack than those based on poly(ether) diols.

EFFECT OF POLYMER MOLECULAR
WEIGHT AND BRANCHING

Poly(ethylene)

It has been observed by a number of investigators that low molecular weight normal paraffins are readily utilized by micro-organisms, [8–10] while their branched isomers are very poorly utilized. We have measured the biodegradability of ten pure linear hydrocarbon samples in the molecular weight range 170 – 620 as shown in Table II. At molecular weights up to and including 451 (C_{32}) a growth rating of 4 was obtained. Above that point a growth rating of 0 was obtained.

We have also measured the biodegradability of several branched hydrocarbons as shown in Table III. All of these samples had molecular weights less than 450. 2, 6, 11-Trimethyldodecane, 2, 6, 11, 15-tetramethylhexadecane and squalane (C_{30}) all gave growth ratings of 0 while their analogous straight chain compounds as shown in Table II gave growth ratings of 4.

In Table IV the effect on biodegradability of varying molecular weight on low and high density poly(ethylene) is reported. Since the high density poly(ethylene) is a linear nonbranched molecule, the molecular weight distribution in samples 1 and 2 is evidently such that the low molecular weight species (below about 500

TABLE I

BIODEGRADABILITY OF COMMERCIAL PLASTICS

Product	Growth rating
1. Poly(ethylene) household wrap	2
2. Sample 1 extracted with toluene	1
3. PVC - epoxidized soybean oil plasticizer	3
4. Sample 3 extracted with toluene	1
5. Poly(propylene)	1
6. Poly(styrene)	1
7. Poly(ethylene terephthalate)	1
8. Poly(vinylidene chloride)	1
9. Acrylonitrile - butadiene - styrene copolymer (ABS)	0
10. ABS - poly(carbonate) blend	0
11. Butadiene - acrylonitrile rubber	0
12. Styrene - acrylonitrile copolymer	0
13. Rubber modified poly(styrene)	0
14. Styrene - butadiene block copolymer	1
15. Poly(methyl methacrylate)	0
16. Rubber modified poly(methyl methacrylate)	0
17. Poly(ethylene terephthalate)	0
18. Poly(cyclohexanedimethanol terephthalate)	0
19. Bisphenol A poly(carbonate)	0
20. Bisphenol A poly(sulfone)	0
21. Poly(4-methyl-1-pentene)	0
22. Poly(isobutylene)	0
23. Chlorosulfonated poly(ethylene)	0
24. Cellulose acetate or butyrate	0
25. Nylon-6, nylon-66, nylon-12	0
26. Poly(urethane)	4
27. Poly(vinyl butyral)	0
28. Poly(formaldehyde)	0
29. Poly(vinyl ethyl ether)	0
30. Poly(vinyl acetate)	1

TABLE II

BIODEGRADABILITY OF STRAIGHT CHAIN HYDROCARBONS

Compound	Formula	Molecular weight	Growth rating
Dodecane	$C_{12}H_{26}$	170	4
Hexadecane	$C_{16}H_{34}$	226	4
Octadecane	$C_{18}H_{38}$	255	4
Docosane	$C_{22}H_{46}$	311	4
Tetracosane	$C_{24}H_{50}$	339	4
Octacosane	$C_{28}H_{58}$	395	4
Dotriacontane	$C_{32}H_{66}$	451	4
Hexatriacontane	$C_{36}H_{74}$	507	0
Tetracontane	$C_{40}H_{82}$	563	0
Tetratetracontane	$C_{44}H_{90}$	620	0

TABLE III

EFFECT OF BRANCHING ON
HYDROCARBON BIODEGRADABILITY

Compound and structure	Molecular weight	Growth rating
2, 6, 11 -Trimethyldodecane, $C_{15}H_{32}$	212	0
2, 6, 11, 15-Tetramethylhexadecane, $C_{20}H_{42}$	283	0
Squalane, $C_{30}H_{62}$	423	0

$$CH_3\underset{\underset{CH_3}{|}}{C}HCH_2CH_2CH_2\underset{\underset{CH_3}{|}}{C}HCH_2CH_2CH_2CH_2\underset{\underset{CH_3}{|}}{C}HCH_3$$

$$(CH_3\underset{\underset{CH_3}{|}}{C}HCH_2CH_2CH_2\underset{\underset{CH_3}{|}}{C}HCH_2CH_2)_2$$

$$(CH_3\underset{\underset{CH_3}{|}}{C}HCH_2CH_2CH_2\underset{\underset{CH_3}{|}}{C}HCH_2CH_2CH_2\underset{\underset{CH_3}{|}}{C}HCH_2CH_2)_2$$

TABLE IV

EFFECT OF POLY(ETHYLENE) MOLECULAR
WEIGHT ON BIODEGRADABILITY

Sample number	Product type	Molecular weight	Growth rating
1	High density poly(ethylene)	10, 970	2
2	High density poly(ethylene)	13, 800	2
3	High density poly(ethylene)	31, 600	0
4	High density poly(ethylene)	52, 500	0
5	High density poly(ethylene)	97, 300	1
6	Low density poly(ethylene)	1, 350	1
7	Low density poly(ethylene)	2, 600	3
8	Low density poly(ethylene)	12, 000	2
9	Low density poly(ethylene)	21, 000	1
10	Low density poly(ethylene)	28, 000	0

molecular weight) are present in sufficient concentration to give
a positive reading in the test. As the average molecular weight
increases into the commercial range, the molecular weight
distribution is shifted so that there is not a sufficient quantity of
these low molecular weight species present to be detected by the
test.

The same conclusion can also be applied to the low density
poly(ethylene) except that samples 6 and 7 appear to be anomalous.
Sample 6 is a grease, which is a very low density highly branched
sample containing very few molecules which are linear. Sample
7 is a crystalline wax with a high density and a higher crystallinity
than sample 6. It, therefore, contains more straight chain molecules below about 500 molecular weight than does sample 6 and
hence received a higher rating.

Samples of high and low density poly(ethylene) have been
thermally degraded using a continuous process developed by Union
Carbide Corporation. [11] Samples were pyrolyzed at temperatures
between 400 and 535° C and were examined for biodegradability with
the results shown in Tables V and VI.

Referring to Table V, it is seen that high density poly-
(ethylene) of 123,000 molecular weight initially, exhibits biodegradability when pyrolyzed to a molecular weight of 3200 or
below. In Table VI, low density pyrolyzed poly(ethylene) shows
biodegradability at a molecular weight of 2100. In both cases the
attack is most likely on those molecules below 500 molecular
weight that are present in the molecular weight distribution.

Poly(styrene)

Laboratory prepared poly(styrene) samples varying in molecular weight from about 600 to about 200,000 give no indication of
being attacked in the agar screening procedure. The data are
shown in Table VII. In addition, over the same range, samples
were prepared with carboxylic acid end groups and carboethoxy
end groups. No enhancement of activity was observed. Poly-
(styrene) was also pyrolyzed down to a molecular weight of 4000

TABLE V

BIODEGRADABILITY OF PYROLYZED
HIGH DENSITY POLY(ETHYLENE)

Pyrolysis temperature, °C	Viscosity average molecular weight	Growth rating
Control	123,000	0
400	16,000	1
450	8,000	1
500	3,200	3
535	1,000	3

TABLE VI

BIODEGRADABILITY OF PYROLYZED
LOW DENSITY POLY(ETHYLENE)

Pyrolysis temperature, °C	Viscosity average molecular weight	Growth rating
Control	56,000	0
400	19,000	1
450	12,000	1
500	2,100	2
535	1,000	3

TABLE VII

BIODEGRADABILITY OF SYNTHESIZED POLY(STYRENE)

Average molecular weight	Growth rating
214,000	0
62,000	0
44,000	0
19,000	0
14,000	0
5,900	0
2,100	0
600	0

TABLE VIII

BIODEGRADABILITY OF PYROLYZED POLY(STYRENE)

Pyrolysis temperature, °C	Average molecular weight	Growth rating
Control	220,000	1
400	93,000	1
450	68,000	0
500	26,000	0
535	4,000	0

as shown in Table VIII. In terms of chain length, a 4000 molecular weight poly(styrene) is comparable to a 1000 molecular weight poly(ethylene). The susceptibility of the pyrolyzed samples to attack by the micro-organisms did not appear to be enhanced.

BIODEGRADABILITY OF RANDOM COPOLYMERS OF ETHYLENE OR STYRENE

A large number of ethylene copolymers of varying composition have been subjected to the screening procedure with no susceptibility to attack being observed for any of the samples. The "comonomers" include vinyl acetate, vinyl alcohol, acrylic acid, sodium or ammonium acrylate, ethyl acrylate, lauryl acrylate, and carbon monoxide. The data are tabulated in Table IX.

In a similar fashion, as shown in Table X, copolymers of styrene with various comonomers containing metabolically active functional groups showed no susceptibility to attack by the test fungi. The comonomers include acrylic acid, sodium acrylate, ethyl acrylate, and dimethyl itaconate.

BIODEGRADABILITY OF POLY(ESTERS)

The only synthetic high polymers which have been found to be biodegradable are those having aliphatic ester linkages in the main chain. Those polymers having aliphatic ester linkages in a pendant position on the main chain, e.g., poly(vinyl acetate), are not utilized by micro-organisms. Previous investigators had also observed the susceptibility of aliphatic poly(esters) to attack by micro-organisms. Table XI lists poly(esters) of varying structure and molecular weight (as measured by reduced viscosity which is proportional to molecular weight), and the growth rating observed for each.

Sample 1, an epsilon caprolactone poly(ester) of about 40,000 molecular weight, which has no branching, is quite readily utilized by fungi and bacteria. Sample 2, a branched poly(ester) derived from pivalolactone, of much lower molecular weight, was not utilized at all.

TABLE IX

BIODEGRADABILITY OF ETHYLENE COPOLYMERS

Comonomer	Percent comonomer	Growth rating
Vinyl acetate	18, 33, 45	1
Vinyl alcohol	30, 70	0
Acrylic acid	15	0
Sodium acrylate	20	0
Ammonium acrylate	20	0
Ethyl acrylate	18	0
Dodecyl acrylate	25	1
Carbon monoxide	6, 48	1

TABLE X

BIODEGRADABILITY OF STYRENE COPOLYMERS

Comonomer	Percent comonomer	Growth rating
Acrylic acid	16	0
Sodium acrylate	16	0
Dimethyl itaconate	30	0
Ethyl acrylate	50	0
Acrylonitrile	28	0
Methacrylonitrile	87	0
Dodecyl acrylate	15	0

TABLE XI

DEGRADABILITY OF POLY(ESTERS)

Sample number	Characterization	Reduced viscosity	Growth rating
1	Caprolactone poly(ester)	0.7	4
2	Pivalolactone poly(ester)	0.1	0
3	Poly(ethylene succinate)	0.24	4
4	Poly(tetramethylene succinate)	0.59	1
5	Poly(tetramethylene succinate)	0.08	4
6	Poly(hexamethylene succinate)	0.91	4
7	Poly(hexamethylene fumarate)	0.25	2
8	Poly(hexamethylene fumarate)	0.78	2
9	Poly(ethylene adipate)	0.13	4
10	Poly(ethylene terephthalate)	high	0
11	Poly(cyclohexanedimethanol terephthalate)	high	0
12	Poly(bisphenol A carbonate)	high	0

TABLE XII

EFFECT OF SOIL BURIAL ON CAPROLACTONE POLY(ESTER)

Months of burial	Tensile strength, P.S.I.	Percent elongation	Weight loss percent
0	2610 ± 103	369 ± 59	0
1.25	1890 ± 215	9 ± 1.4	-
2.0	1610 ± 180	7 ± 2.0	8
4.0	520 ± 220	2.6 ± 1.1	16
6.0	100	negligible	25
12.0	negligible	negligible	42

Poly(esters) based on fumaric acid, which is an unsaturated, dibasic acid, appear to be utilized more poorly than those based on saturated dibasic acids such as succinic and adipic acid.

A marked dependence of biodegradability on molecular weight was observed for poly(tetramethylene succinate) (samples 4 and 5). Aromatic structures as exemplified by samples 10, 11, and 12 render the poly(ester) unassimilable.

The poly(ester) derived from the ring opening polymerization of epsilon-caprolactone was chosen for further biodegradability testing by the soil burial technique. A polymer sample of about 40,000 molecular weight was molded into tensile test bars, which were found to have an ultimate tensile strength of 2610 P.S.I., and an ultimate elongation of 360%, measured at room temperature. Test bars of this material were buried in soil and removed at intervals of 1.25, 2.0, 4.0, 6.0, and 12.0 months for testing. With increasing length of soil burial, the test bars became more pitted and eroded and were made weaker as shown by the data in Table XII. At the end of twelve months the samples were too weak for measurement of strength properties and had lost 42% of their original weight.

Scanning electron micrographs of the surface of the two month soil-buried sample reveal the extent of the attack. Figure 1 is a photograph of the surface before any degradation, magnified about 980 times. The streaks and straight lines are replicas of the surface of the mold in which the tensile bar was made. Figure 2 shows the surface of a tensile bar at about the same magnification (950 x) that has been soil-buried for two months. The deep pitting, channeling and cavernous appearance resulting from the degradation process is readily apparent in these pictures.

Small containers were also injection molded from the caprolactone poly(ester) and buried in the same soil mixture. Examination for weight loss after 2, 4, 6, and 12 months revealed the data shown in Table XIII. The weight loss is greater here than in the tensile bars because of the greater surface area of these samples. Figure 3 shows the progressive degradation of the sample.

TABLE XIII

CAPROLACTONE POLY(ESTER):
SOIL BURIAL SMALL CONTAINERS

Months of burial	Weight loss percent
0	0
2	12
4	29
6	48
12	95

CONCLUSIONS

Aliphatic poly(esters) and derivatives were the only synthetic high molecular weight polymers found to be biodegradable in a study embracing the large volume thermoplastic packaging plastics as well as many experimental polymers. Although high molecular weight poly(ethylene) is not biodegradable, pure linear paraffin molecules below about 500 molecular weight were found to be utilized by micro-organisms. Low molecular weight poly(ethylenes) obtained by synthesis or pyrolysis also supported microbial growth. Low molecular weight poly(styrene) (600 mol. wt.) did not support growth of micro-organisms in our tests. Placement of metabolically active groups at the chain ends or, through copolymerization, along the chain did not observably enhance biodegradability.

The onset of biodegradability below 500 molecular weight observed for straight chain alkanes suggests that the enzymes which catalyze the degradation of normal paraffins through β-oxidation are unable to complex with the chain ends in the higher molecular weight materials. These ends are both in low concentration and may be relatively inaccessible due to the folded chain configuration of the polymer molecule. The ease with which high molecular weight epsilon-caprolactone poly(ester) is assimilated suggests that extra-cellular enzymes are available which are capable of cleaving this high polymer, prior to its assimilation by the organism.

FIGURE 1. Scanning electron micrograph of caprolactone poly(ester): not soil-buried.

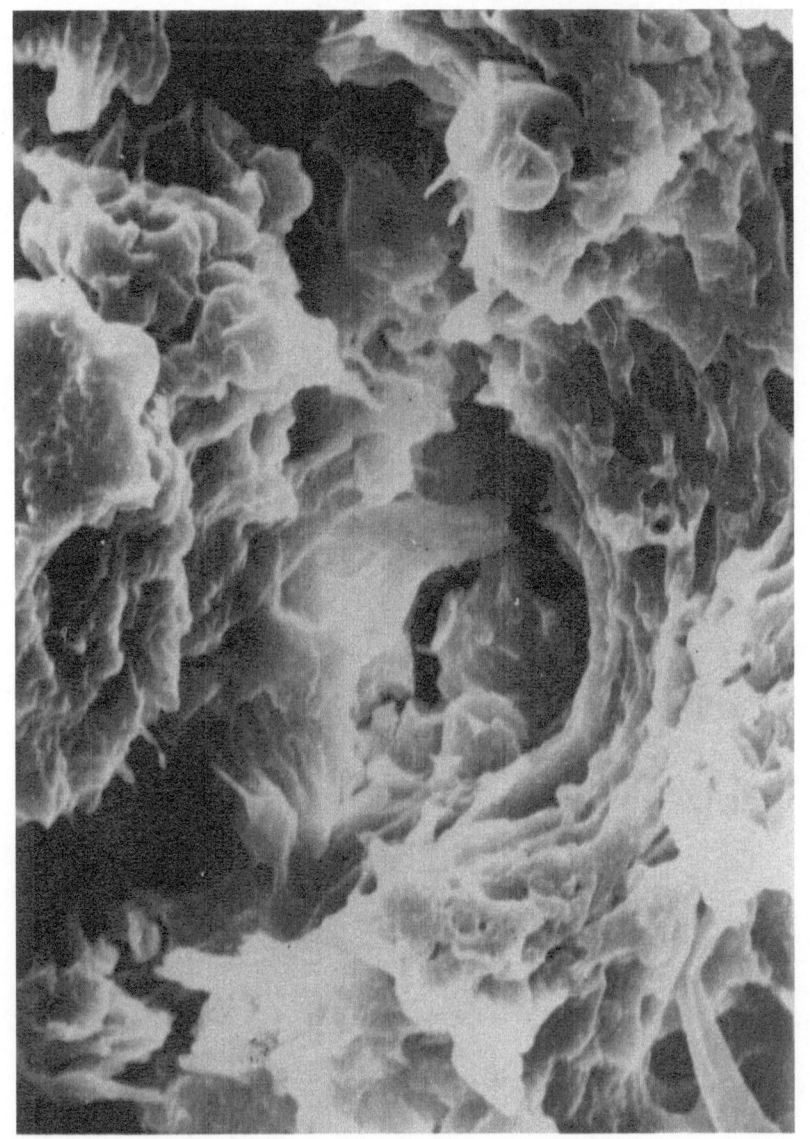

FIGURE 2. Scanning electron micrograph of caprolactone poly(ester): two months soil-buried.

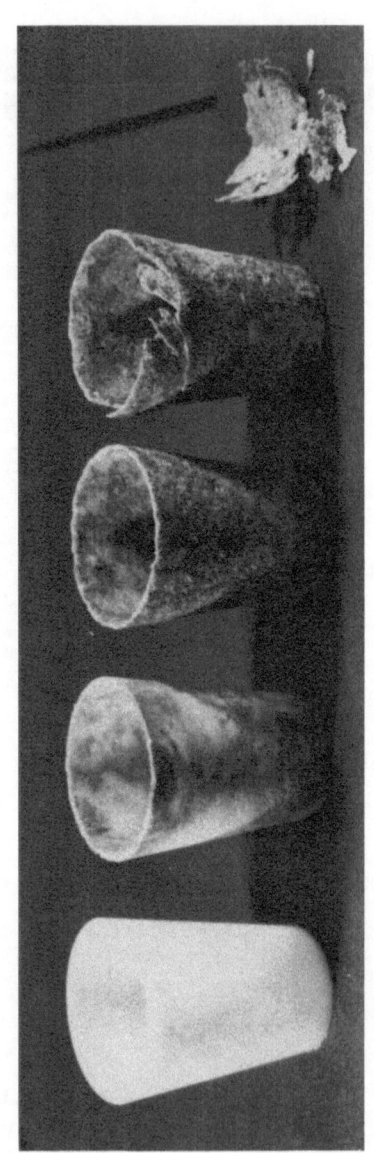

FIGURE 3. Caprolactone poly(ester) small containers: 0, 2, 4, 6, and 12 months soil-buried.

ACKNOWLEDGMENTS

This study was partially funded by the Office of Solid Waste Management, Environmental Protection Agency, through a research and development contract (CPE-70-124). The authors wish to acknowledge the advice and encouragement offered by the Contract Monitor, Mr. Clarence Clemons, of the Agency. They also wish to acknowledge the excellent laboratory support provided by Mr. Kermit Craig and Mr. Jesse Doll of this laboratory.

REFERENCES

1. M. E. Fulmer and R. F. Testin, "Role of Plastics in Solid Waste", Battelle Memorial Institute, 1967.

2. H. J. Hueck, Plastics, October, 1960, p. 419.

3. C. J. Wassell, SPE Trans., July, 1964, p. 193.

4. F. Rodriques, Chem. Tech. July, 1971, p. 409.

5. G. Tirpack, SPE J., 26, 26 (1970).

6. R. T. Darby and A. M. Kaplan, Appl. Microbiol., 16, 900 (1968).

7. L. Jen-Hao and A. Schwartz, Kunstoffe, 51, 317 (1961).

8. P. K. Barua et al., Appl Microbiol., 20, 657 (1970).

9. E. Merdinger and R. P. Merdinger, Appl. Microbiol., 20, 561 (1970).

10. T. L. Miller and M. J. Johnson, Biotechnol. Bioeng., 8, 567 (1966).

11. J. E. Potts, Ind. Water Eng., 7, 32 (1970).

EFFLUENTS FROM THE MUNICIPAL
INCINERATION OF PLASTICS

Richard B. Engdahl, Horatio H. Krause, and
Paul D. Miller

Battelle's Columbus Laboratories
505 King Avenue, Columbus, Ohio, 43201

INTRODUCTION

Incineration is an ancient method for disposal of the combustible portion of community wastes which includes a small but increasing proportion of plastics. Too often incineration is also a source of annoying air pollution because of poor combustion or because some of the gaseous, liquid or solid combustion products are carried off in the exhaust gases. This does not have to be. In the United States, excess emissions from incinerators have been tolerated because we have tended "to apologize that waste disposal costs anything at all". And since incineration, even without good pollution control, costs considerably more than most sanitary landfills, the idea of adding to that cost in order to add equipment to clean up the emissions to the atmosphere has been slow to come to the U. S. However, the emergence of many large successful power-generating incinerators abroad, with elaborate, high-efficiency electrostatic precipitators to capture over 95% of the flyash, has demonstrated that much poorer countries than the U. S. have decided that they want and can afford clean incinerators. Now these are starting to be built by a few municipalities in the U. S. Meanwhile, over 75% of municipal solid waste in the U. S. still goes to open dumps and landfills.

Only 1 to 3% of the municipally collected waste in the U. S. is plastic,[1] and less than 10% of the 200 million tons of waste

81

collected each year are incinerated. Hence only a small fraction
of the plastic discarded every year is incinerated. Nevertheless
this constitutes 0.3 to 0.6 million tons of plastic burned each year,
with increasing amounts to come in the future. Of the present
total about 31% (100,000 to 200,000 tons) is poly(vinyl chloride).[2]

INCINERATOR CHARACTERISTICS

Many varieties of incinerators are in use, predominantly
fixed or moving grate types which were evolved from coal-burning
practice. Some rotary kiln types are also used.[3] In most cases
the major types have demonstrated their ability to burn hetero-
geneous community wastes smokelessly. In the predominant fixed
bed or oscillating or moving bed types, primary air forced upward
through the slowly agitated bed gasifies and partially burns the
combustibles. A major portion of the excess combustion air is
then injected at high velocity through multiple jets located above
the burning bed to provide ample oxidation and intense mixing to
complete the burning of the volatile gases rising from the bed.

Incineration of Plastics

The rising quantities of plastics in municipal wastes has not
overtaxed the ability of well designed, well maintained, and well
operated incinerators to burn the plastics and associated waste
without smoke. In fact the high heat value of most plastics has
contributed, along with the rising paper content of U. S. wastes,
to steadily increasing heat value of the wastes, thus reducing the
problem of low temperature operation during seasons when moisture content is high.

Laboratory measurements of combustion under conditions
of a deficiency of air show that soot and distillation products can
be evolved from any combustible, plastic or otherwise, but many
tests of operating incinerators attest to the fact that such emissions
are not produced during good operation. In general, the smaller
the incinerator, the more likely operation is to be a part-time job
with consequent neglect and periods of smoky operation.

Characteristically incinerators produce lower concentrations of nitrogen oxides than do boiler furnaces or engines, probably because combustion temperatures are usually lower. The introduction of plastics appears to raise NO_x emissions slightly because of somewhat higher temperatures.

Boettner[4] and associates at the University of Michigan have carried out extensive studies on pyrolytic products of plastics. Plastics examined include:

Poly(vinyl chloride)
a. four polymers
b. three PVC plastics

Poly(phenylene oxide)
a. four types

Polyimide

In work with PVC they measured the amounts of over 20 combustion products. Table I shows the amounts of the eight most abundant gases from six types of PVC including floor tile and wire insulation formulations. It can be seen that major constituents are CO_2, HCl, and CO. Significant amounts of benzene, methane, ethane, ethylene, and toluene are also evolved. The HCl is released at temperatures near 300° C.

The major combustion products from polyimides are listed in Table II.

Tsuchiya and Sumi[5] studied the thermal decomposition of PVC under simulated uncontrolled building fires. Although the results are not directly applicable to incineration conditions, they indicate the effect of temperature and inert atmosphere on products. Table III shows their results. Neither chlorine nor phosgene could be detected in these rather extreme conditions.

Ball[6] found that CO_2 and CO were the major volatile products from heating poly(phenylene oxide). Methane, toluene, ethylbenzene, and styrene are next most abundant.

TABLE I

MAJOR PYROLYSIS PRODUCTS OF POLY(VINYL CHLORIDE)

	A		B		C	
	PVC, MW 111,000	Wire insulation containing 51% PVC (A)	PVC, MW 250,000	Wire insulation containing 57% PVC (B)	85:15 copolymer vinyl chloride/ vinyl acetate MW 55,000	Floor tile containing 35% C
			Amount as mg/g			
CO_2	729.	616.	730.	1182.	923.	456.
HCl	583.	273.	584.	333.	500.	73
CO	442.	67.	403.	90.	292.	31.
Benzene	36.	10.	29.	11.	28.	0.86
Methane	4.6	6.6	5.8	6.8	4.4	0.30
Ethane	2.2	3.0	2.5	2.9	2.3	0.13
Ethylene	0.58	2.3	0.33	2.0	0.60	0.13
Toluene	1.3	0.94	1.1	1.0	0.96	0.04
Acetic acid	--	--	--	--	96.	20.

TABLE II

MAJOR PYROLYSIS PRODUCTS OF POLYIMIDE

Compound	Amount mg/g
CO_2	1600 - 2000
CO	350 - 575
H_2O	> 50
NH_3	5 - 50
Nitrogen oxides	5 - 50
HCN	15

Cornish[7] indicated that the major combustion products from phenolic, melamine, nitrocellulose, and poly(ethylene) plastics were as follows:

Phenolic and melamine	CO_2, CO, HCN, NH_3
Nitrocellulose	CO, oxides of nitrogen
Poly(ethylene)	CO_2, H_2O, vaporized poly(ethylene)

Of the products mentioned in the preceding paragraphs, the one of greatest concern from the standpoint of emissions or of deterioration of incineration equipment is HCl. The work of Carotti and others has demonstrated that a ton or more of this acidic substance can be evolved from a normal sized incinerator in one day.

Another deleterious chemical released from some plastics and from rubbers is SO_2.

Both HCl and SO_2 as well as organic acids form strongly acidic solutions when wet scrubbers are used in incinerators for pollution control, particularly if the water is recirculated. In this case the water soon drops to a pH of 2.0 or lower. Data furnished by the operators of the Montgomery County, Ohio, incinerators to

TABLE III

DECOMPOSITION PRODUCTS OF POLY(VINYL CHLORIDE)

Temp. of decomp., °C	350	600	850	350	600	850	350	600	850
Weight of sample, g	0.5	0.5	0.5	0.5	0.5	0.5	0.25	0.25	0.25
Atmosphere	He	He	He	Air	Air	Air	Air	Air	Air
Decomposition products, weight-percent sample									
HCl	53.2	55.5	57.9	47.2	46.0	40.6	48.5	42.3	24.3
CO	--	--	--	1.0	35.6	16.6	1.2	43.0	18.5
CO_2	--	--	--	1.7	40.8	54.8	2.5	65.7	99.7
H_2	--	0.06	0.47	--	0.16	0.50	--	0.11	0.37
CH_4	--	1.0	3.2	--	1.7	3.4	--	1.7	2.4
C_2H_6	--	0.69	0.32	--	0.14	0.06	--	0.14	0.03
C_2H_4	--	0.52	2.5	--	0.63	1.3	--	0.38	0.41
Benzene	5.9	5.6	5.9	5.4	4.7	3.1	5.1	4.8	1.3
Toluene	0.05	0.67	0.87	--	0.22	0.89	--	0.18	0.31
Residue	37.4	6.3	5.1	40.2	2.4	--	39.5	0.3	--

Battelle-Columbus during a PHS grant study are shown in Figure
1. This figure shows the buildup of various constituents in the
scrubber water at the South Plant over a period of about two weeks.
It can be seen that the pH remained quite low (near 2.5) and that
the chloride rose from about 9,000 ppm to 40,000 ppm. It is estimated that the buildup of chloride in the circulating solution in the
example cited above is about 500 lb/day.

Other factors leading to high corrosion in incinerator scrubbers are elevated temperatures (140-190° F), abundant oxygen
supply, and high gas velocities.

It should not be inferred, however, that all the HCl coming
from an incinerator is due to the combustion of plastics. It has
been estimated that about half is coming from that source, but
even this is an appreciable amount.

Kaiser and Carotti[8] reviewed Swedish reports on experiments in 1969 at the Fagersta incinerator burning refuse at rates
of 3580 to 5000 lbs per hour containing 2.2% PVC. In two tests,
60 to 64.9% of the theoretical chlorine appeared as HCl in the gases.
Adding lime reduced the HCl emitted to 45% of the theoretical.

Fuller et al. obtained U. S. Patent 3,556,024 on January 19,
1971, on a "Method of Reducing Halogen Emissions from the Incineration of Halogen-Containing Plastics" by the addition of finely
divided alkali. Seventy-five percent reduction of HCl emission was
claimed.

In tests at Babylon, Long Island, Kaiser and Carotti added
various plastics to municipal refuse. The only appreciable effect
on emissions was from the addition of PVC. Without any addition
HCl emission was 511 ppm. With 2% PVC added, emission rose
to 1990 ppm and with 4% PVC added, it rose to 3030 ppm.

It should be pointed out that the burning techniques used in
the incineration of plastics are of great importance. Most plastics
have a high heat value so that burning rate must be controlled. On
the other hand, too slow burning can cause excessive smoke formation and CO evolution. Burning should also be at such a rate that

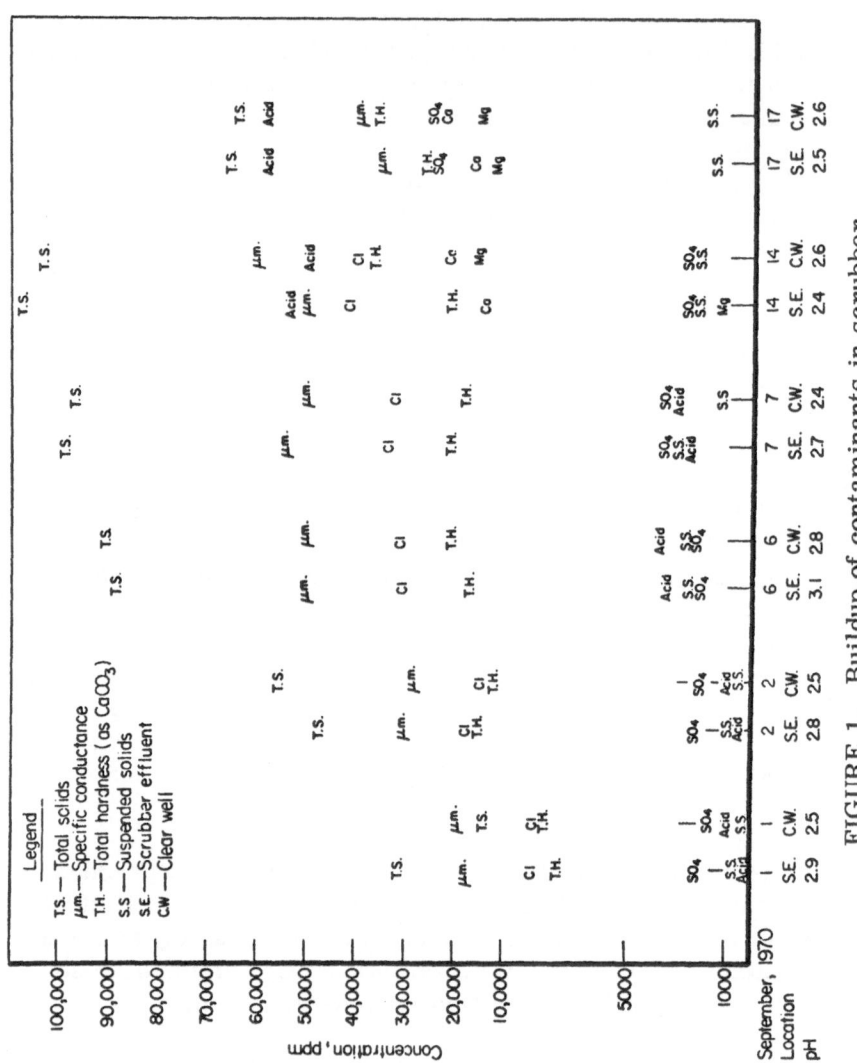

FIGURE 1. Buildup of contaminants in scrubber water of Montgomery County, Ohio, incinerator.

that melted plastic does not flow into the grate openings, solidify from the incoming air, and block the agitating motion of the grate.

FIRE RESISTANT PLASTICS

For reasons of fire safely, research is underway to increase the fire resistance of plastics by the introduction of halogens and phosphorus.[9] Widespread appearance of such plastics in incinerators could introduce new decomposition products.

Another current source of particle emission from incinerators is the finely divided inert filler in coated papers and in some plastics. If lime were added to PVC to capture HCl, it might increase dust emission.

SUMMARY

In terms of current incinerator technology, burning of plastics in municipal incinerators need not constitute a source of air pollution. However, improper or careless operation can lead to a smoky effluent. The contribution of plastics to the particulate loading during efficient combustion is negligible.

No appreciable amounts of highly toxic gases have been detected in the effluents from municipal incinerators even when large amounts of plastics were added to the charges. HCl from poly(vinyl chloride) is the chief effluent of concern at the present time. No standards for HCl emissions have been set as yet by national or local agencies. The effluent levels currently may range from a few ppm to several hundred ppm depending on whether an efficient wet scrubber is utilized in the incinerator. The greatest problem arising from the HCl is the corrosion of the incinerator components.

REFERENCES

1.   Plastics in Solid Waste, National Industrial Pollution Control Council, Sub-Council Report, March 1971, p. 9. For sale by Supt. of Documents, U. S. Govt. Printing Office, Washington, D. C. 20402, 25 cents.

2.   M. E. Fulmer and R. F. Testin, Report on the Role of Plastics in Solid Waste, 1968, Society of the Plastics Industry, Inc., 250 Park Avenue, New York, N. Y. 10017.

3.   R. C. Corey, editor, Principles and Practice of Incineration, Wiley, New York, 1970.

4.   E. A. Boettner, G. Ball, and B. Weiss, J. Appl. Polym. Sci., 13, 377 (1969).

5.   Y. Tsuchiya and K. Sumi, J. Appl. Chem., 17, 364 (1967).

6.   G. Ball, B. Weiss, and E. A. Boettner, Amer. Ind. Hyg. Assn. J., 31, Sept. 1970.

7.   H. H. Cornish, "Toxicological Aspects of Flammability and Combustion", Proc. Univ. Detroit Polymer Conf., June 1970.

8.   A. A. Carotti and E. R. Kaiser, "Concentration of 20 Gaseous Chemical Species in the Flue Gas of a Municipal Incinerator", Paper no. 71-67, 64th Annual Meeting, APCA, Atlantic City, June 1971. See also E. R. Kaiser and A. A. Carotti, "Municipal Incineration of Refuse with 2 Percent and 4 Percent Additions of Four Plastics", Report to the Society of the Plastics Industry, Inc., 250 Park Avenue, New York, N. Y. 10017.

9.   C. J. Hilado, Chemtech, 2, 232 (1972).

BIBLIOGRAPHY

M. M. O'Mara, L. B. Crider, and R. L. Daniel, "Combustion Products from Vinyl Chloride Monomer", Amer. Ind. Hyg., 32, 153 (1971).

Editor, "Can Plastics be Incinerated Safely", Environ. Sci. and Tech., 5, 667 (1971).

W. R. Fuller and B. H. Bieler, "Method of Reducing Halogen Emissions from the Incineration of Halogen-Containing Plastics", U. S. Patent 3,556,024, Jan. 19, 1971.

G. P. Fong, A. C. Mack, and J. N. McDonald, "Role of Plastics in Solid Waste -- A Status Review", Proc. Inst. Solid Waste, Framingham, Mass., 1970.

E. A. Boettner, G. Ball, and B. Weiss, "Analysis of the Volatile Combustion Products of Vinyl Plastics", J. Appl. Polym. Sci., 13, 377 (1969).

G. Ball, B. Weiss, and E. A. Boettner, "Analysis of the Volatile Combustion Products of Polyphenylene Oxide Plastics", Amer. Ind. Hyg. Assn. J., 31, 572 (1970).

A. V. Bjorkman, "Plastics Needn't be a Problem", The American City, pp. 148-151, Sept. 1970.

E. H. Coleman and C. H. Thomas, "The Products of Combustion of Chlorinated Plastics", J. Appl. Chem., 4, 379 (1954).

B. Ranby, "Technical Environmental Data on Plastics", Fasfat-bolaget, Sweden, 1969.

E. Schmitz and E. Wogrally, "Test Results on Corrosion Caused by the Combustion of Macromolecular Materials", Allgemeine und Praktische Chemie, 20, 251 (1969).

R. Heimburg, "Environmental Effects of the Incineration of Plastics", AIChE Meeting, Houston, Texas, Feb.-Mar., 1971.

SKYDD-69, "International Symposium on Corrosion Risks in Connection with Fire in Plastics", Swedish Fire Protection Assn.,
Oct., 1969, Stockholm.

A. A. Carotti and R. A. Smith, "Air Borne Emissions from Municipal Incinerators", Bureau Solid Waste Management, H. E. W.,
July, 1969.

S. Brohult and H. Gralen, "Plastics from an Environmental Standpoint", Royal Swedish Academy of Engineering Sciences, Stockholm, 1969.

G. Ball and E. A. Boettner, "Combustion Products of Plastics
and their Contribution to Incineration Problems", talk to Amer.
Chem. Soc., Chicago, Sept. 17, 1970.

P. D. Miller and H. R. Krause, "Factors Affecting the Corrosion
of Boiler Steels in Municipal Incinerators", Corrosion, 27, 31 (1971).

J. Tichatschke, "Investigations of Emissions of Refuse Incinerating Installations", Mitteilungen V G B, 51, 219 (1971).

K. Fassler, H. Leil, and H. Spahn, "Corrosion in Refuse Incineration Plants", Mitteilungen, 48, 126 (1968).

C. W. Teller and H. Bohne, "Chlorwasserstoff in Rauchgasen von
Müllverbrennungsanlagen", Müll-Abfall-Abwasser, 10, 28 (1969).

Battelle Research Outlook, "What's Ahead in Solid Waste Management", 1971.

THE AUTO-IGNITION OF MULTICOMPONENT FIBER SYSTEMS

Bernard Miller, J. Ronald Martin, and
Charles H. Meiser, Jr.

Textile Research Institute
Princeton, New Jersey, 08540

ABSTRACT

Using a simple experimental procedure for determining the
time until ignition after a material is plunged into a heated air
environment, it is possible to obtain kinetic data for this auto-
ignition process. Results on homogeneous polymeric materials
show that, while the rate controlling factors in every case are
physical rather than chemical, ignition times at any single temperature depend on the basic thermal degradation process of each
polymer. The study also includes work on multicomponent systems,
including a mixed fiber blend as well as polymers treated with non-
polymeric additives. For each combination studied it is possible
to detect whether the presence of a second component has had only
a physical influence on ignition, or whether thermal degradation
has been affected by chemical interaction. Such data can serve to
distinguish modifiers of flammability that act on the ignition process from those that are literally flame retardants or accelerators.

INTRODUCTION

Auto-ignition is defined as the generation of flaming combustion resulting from placing a material in contact with heated air
in the absence of any spark or flame. The process depends on

93

convective heat transfer and is a highly reproducible phenomenon, in contrast to ignition by radiant heating, flame impingement or conductive heat transfer methods. Techniques for obtaining and analyzing auto-ignition data have been developed[1] and are herewith applied to studying how the thermal and combustion behaviors of fiber-forming polymers are modified by the presence of various additives, including another polymer. Originally, interest in this work stemmed from its relevance to problems anticipated during incineration of such materials. However, it has been found that results obtained can be used also to reveal fundamental information about thermal decomposition and flame retardancy.

EXPERIMENTAL

Apparatus

The experimental apparatus designed for studying auto-ignition is shown schematically in Figure 1. Its essential features are:

1. A vertical, cylindrical, ceramic furnace (2-1/4 in diameter by 10 in deep) with temperature control and a fixed thermocouple to monitor the temperature of the air at a point near to where the sample is placed.

2. A removable sample holder at the end of which are attached two prongs onto which the sample is impaled. A thermocouple is stationed between the sample prongs near the sample but not touching it.

3. A strip chart recorder which monitors the output of the thermocouple on the sample holder.

Samples

Samples used in these auto-ignition studies were single and multiple layers of fabric (1 in x 1/2 in) stapled onto a wire screen having the same dimensions. For thermoplastic materials the screen serves to preserve the geometrical integrity of the sample

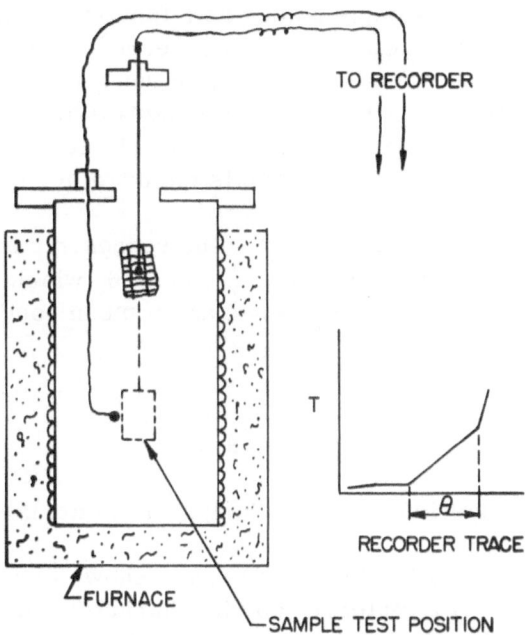

TO RECORDER

RECORDER TRACE

FURNACE

SAMPLE TEST POSITION

FIGURE 1. Apparatus for auto-ignition studies.

during the melting which precedes ignition. It also functions as
an inert heat sink which acts in direct competition with the sample
and lengthens slightly the time to ignition. Therefore non-thermo-
plastic materials were also stapled to a screen so that all materials
might be compared on the same basis.

Procedure

    The sample is impaled onto the two prongs of the sample
holder and the thermocouple positioned close to the sample but
not touching it. The furnace temperature is adjusted to the
desired level and the sample is then placed into the furnace by
quickly thrusting the sample holder into position. The time
required to place the sample into position is of the order of a
fraction of a second -- short enough to give a sharp inflection
point on the recorder which measures the output of the

thermocouple on the sample holder as a function of time. The
moment of ignition is represented by a second inflection point in
the recorder output when the combustion reaction(s) abruptly increase (flaming). The time between these two inflection points
is termed the ignition time, θ , measured in seconds. A typical
readout of the strip chart recorder is included in Figure 1.

Each reported experimental value represents a mean average
of five to ten independent determinations of θ, which in general
show a standard deviation of 0.5 sec and a mean coefficient of
variation of 5% or less.

RESULTS

Auto-Ignition of Single Component Materials

Previously reported studies[1] have shown that the auto-
ignition behavior of a material may be characterized by the linear
relationship between observed ignition time, θ , and sample mass
(at constant sample area) for each of a series of oven temperatures.
The extrapolation of this linear relationship to zero mass produces
an intrinsic ignition time, θ_0, which describes the response of an
infinitely thin sample at the given oven temperature. Data of this
type have been obtained for a wide variety of materials. The auto-
ignition behavior of cotton fabric, as shown in Figure 2, is typical.

The temperature dependence of such intrinsic ignition times
is shown in Figure 3 in the form of Arrhenius plots (log $1/\theta_0$ vs.
$1/T$) for all the materials investigated up to now. The relationship is consistently linear with an apparent activation energy
(obtained from the slopes) corresponding to 8-10 kcal/mole, except for Nomex Ⓡ which shows an activation energy of 17 kcal/mole.
These low activation energies coupled with the fact that the same
activation energy is obtained for nearly all chemical species
investigated suggest that this ignition phenomenon is controlled by
physical processes (heat or mass transfer) which would be associated with such a low activation energy. The activation energy
is derived using the temperature of the air in the oven and therefore
probably represents the kinetics in the gas phase where diffusion of
combustible gases from the sample would be expected to be important.

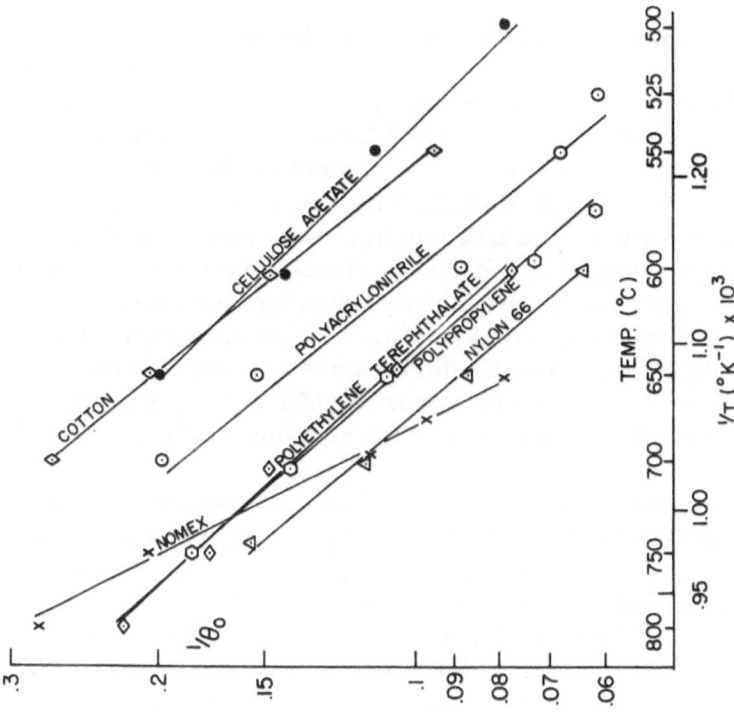

FIGURE 3. Arrhenius plots of intrinsic ignition time (θ_0) as a function of furnace air temperature.

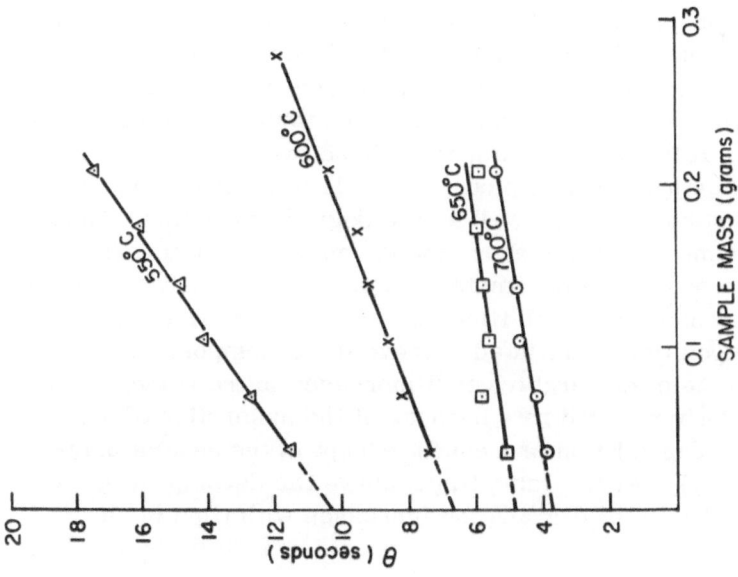

FIGURE 2. Auto-ignition behavior of cotton fabric.

Auto-Ignition of Flame Retardant Materials

Auto-ignition studies were then extended to fibers treated with flame retardants. Figures 4 and 5 show the auto-ignition behavior of cottons treated with 4% diammonium phosphate (DAP) and 5% 70/30 Borax/Boric Acid (B/BA), respectively, in the form of relationships between observed ignition time and mass for a series of furnace temperatures. The extrapolations of these linear relationships to zero mass produce intrinsic ignition times, θ_0. For purposes of comparison, the behavior of an untreated cotton is shown by light dashed lines. (No data are reported for un-treated cotton at a furnace temperature of 500° C since inconsis-tent ignition was observed at temperatures below 550° C.)

Figure 4 shows that, for a given furnace temperature, the addition of 4% DAP to cotton decreases the intrinsic ignition time and decreases the slope of the ignition time-sample mass rela-tionship slightly (but consistently). Figure 5 shows that for a given furnace temperature the addition of 5% B/BA decreases the ignition time to about the same level as does the 4% DAP but in-creases the slope of the ignition time−sample mass relationship.

This contrasting behavior can best be explained by consider-ing that the observed ignition time is made up of two sequential steps: (1) heating the material up to its decomposition tempera-ture, and (2) the chemical degradation processes that start at the decomposition temperature and lead to the formation of combustible gases. The relative contributions of these two processes are shown in Figure 6 as hypothetical plots of sample temperature vs. time (solid and broken lines) as well as the degree of chemical degra-dation vs. time (solid line above the decomposition temperature) with sample mass as a parameter. Before decomposition, sample temperature increases with time in a decaying exponential manner until decomposition is initiated. Above the decomposition tempera-ture, the sample temperature still increases in the same manner, but at a slightly reduced rate because of the absorption of heat by endothermic degradation processes (except in the case of poly-(acrylonitrile)). At the same time, above the decomposition temp-erature the degree of degradation increases with time in an

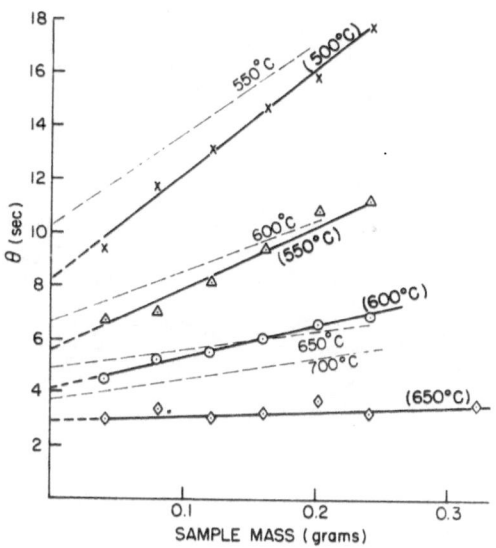

FIGURE 4. Auto-ignition of untreated and flame retardant cotton. Solid lines = cotton + 4% DAP; light broken lines = untreated cotton.

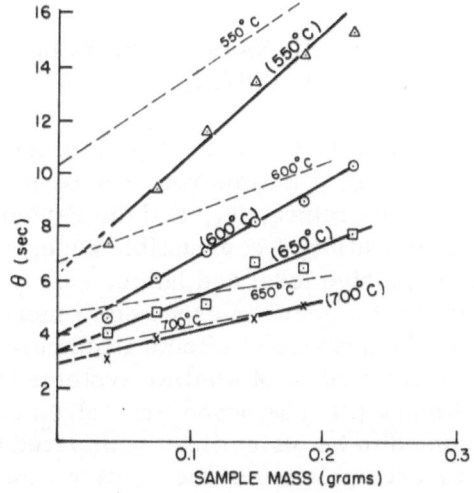

FIGURE 5. Auto-ignition of untreated and flame retardant cotton. Solid lines = cotton + 5% (70/30) B/BA; light broken lines = untreated cotton.

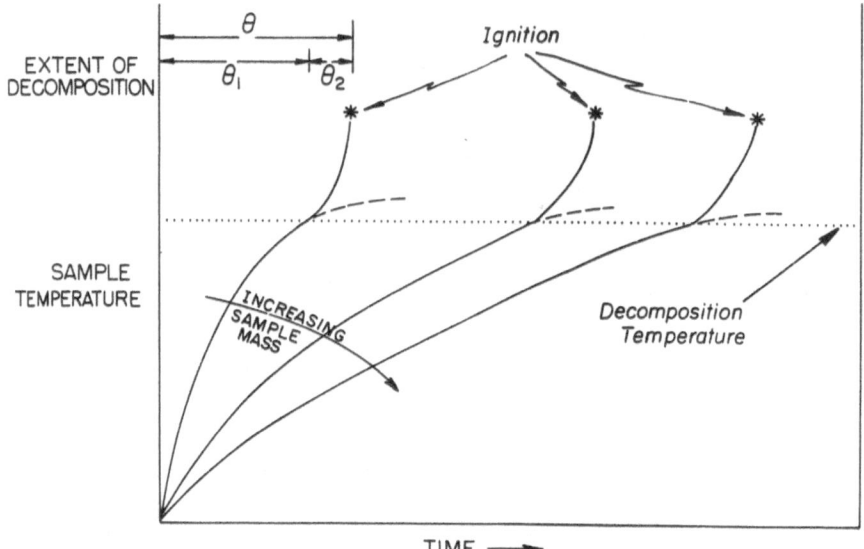

FIGURE 6.   Generalized hypothetical rate process analysis for
auto-ignition.

accelerated exponential manner (as the result of increasing sample
temperature) up to the point of ignition.

        As illustrated in Figure 6, the total time to ignition, θ,
should consist of two parts: the time required to heat the sample
to its decomposition temperature, θ_1, and the remaining time
required for the concentration of combustible gases around the
sample to reach an explosive level and ignite, θ_2. From heat
transfer considerations θ_1 would be expected to increase linearly
with sample mass in the absence of effects from bulk thermal con-
ductivity which in model studies of similar systems have been
shown to be insignificant. [2] The dependence of θ_2 on sample
mass has been assumed to be insignificant compared to that of
θ_1. Thus the linear extrapolation of the ignition time–sample
mass relationship to zero mass eliminates θ_1 and results in an
intrinsic ignition time, θ_0, as if the sample had been instantaneously
heated to its decomposition temperature. Therefore, at a given
temperature, θ_0 will be equal to θ_2. This argument is illustrated
in Figure 7.

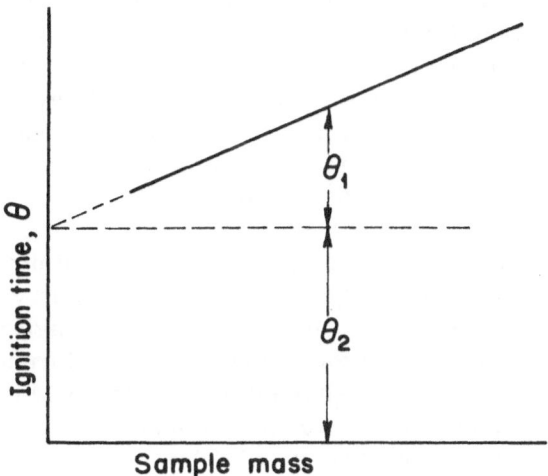

Figure 7.   Resolution of observed ignition time into its component parts.

Since θ_0 is controlled by the decomposition processes of the
material, the decreased intrinsic ignition times brought about by
addition of either DAP or B/BA to cotton imply modification, kinetic
or otherwise, of the cellulose decomposition process. It does not
necessarily mean that the decomposition temperature has changed
since the intrinsic ignition time is independent of this temperature.
Previously reported thermogravimetric analysis studies[3] have
shown that the addition of 4% DAP lowers the decomposition temperature of cotton approximately 50° C while the addition of 10%
B/BA has no affect on this property. However, the results presented here show that, despite such contrasting behavior, both
additives affect the decomposition mechanism as is evident from
the decreased intrinsic ignition times. The increased slope of the
B/BA lines with respect to untreated cotton reflect an increased
dependence on mass for the actual time required to heat a finite
sample to its decomposition temperature. This indicates slower
heating of the material, probably from a decrease in thermal
absorptivity. This behavior may perhaps partially be explained
by the fact that borax has a higher specific heat (0.385 cal/g/° C)
than cellulose (0.32 cal/g/° C). An alternate explanation for the

increased slopes in the B/BA treated cotton is that the presence
of B/BA may increase the equilibrium moisture content of the
cotton so that the time required to heat the material to its decomposition temperature would be lengthened by the energy requirement for water vaporization. Any endothermic reactions of the
additive could also be expected to lengthen the time required to
heat the material to its decomposition temperature.

Figure 8 compares the auto-ignition behavior of untreated
nylon 66 with the same material made flame retardant by treatment with 3.75% thiourea. The effect of thiourea on nylon is
analogous to that of DAP on cotton in that at a given furnace temperature the lines are shifted to shorter ignition times with little
or no reduction in slope. These results indicate that thiourea
affects the decomposition mechanism of nylon.

Figure 9 compares the auto-ignition behavior of untreated
poly(ethylene terephthalate) (PET) and PET treated with 12%
tris-2,3-dibromopropyl phosphate (T23P). The data at each
furnace temperature show the same intrinsic ignition time for both
treated and untreated PET with differences in ignition times between the treated and untreated materials being observed only at
finite masses. These results indicate that T23P does not affect
the decomposition of PET and that prior to ignition T23P is present
as an inert substance which alters the observed ignition time only
through mass effects. The fact that the slopes are decreased by
addition of T23P indicates that the thermal absorptivity of the untreated material is less than that of the treated material.

By way of contrast, Figure 10 shows the effect of 4.8% DAP
on the auto-ignition behavior of PET. The effect is analogous to
that of DAP on cotton in that for a given furnace temperature the
lines are shifted to shorter intrinsic ignition times with no significant change in the slope. These results indicate that DAP affects
the decomposition mechanism of PET.

For purposes of comparison, the intrinsic ignition times for
these flame retardant materials (along with the pure untreated
materials) are collected in Figure 11 in the form of Arrhenius
plots (log $1/\theta_0$ vs. $1/T$). The slopes of the lines for all materials
fall into the previously reported range of 8 - 10 kcal/mole.

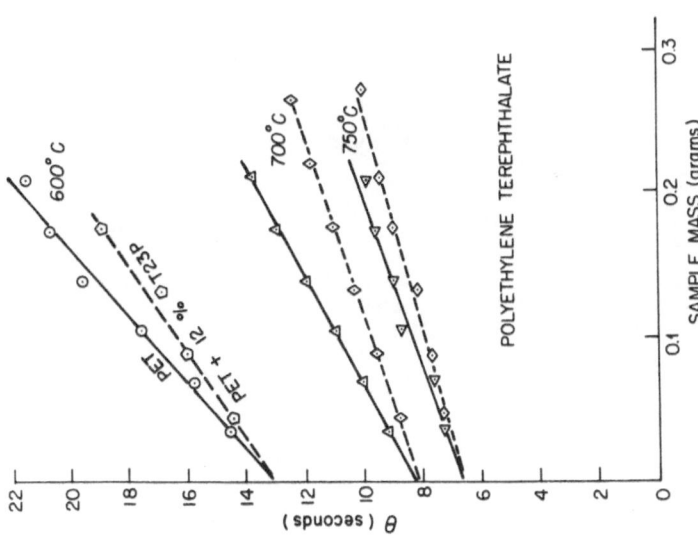

FIGURE 9. Auto-ignition of untreated and flame retardant poly(ethylene terephthalate). Solid lines = untreated PET; broken lines = PET + 12% T23P.

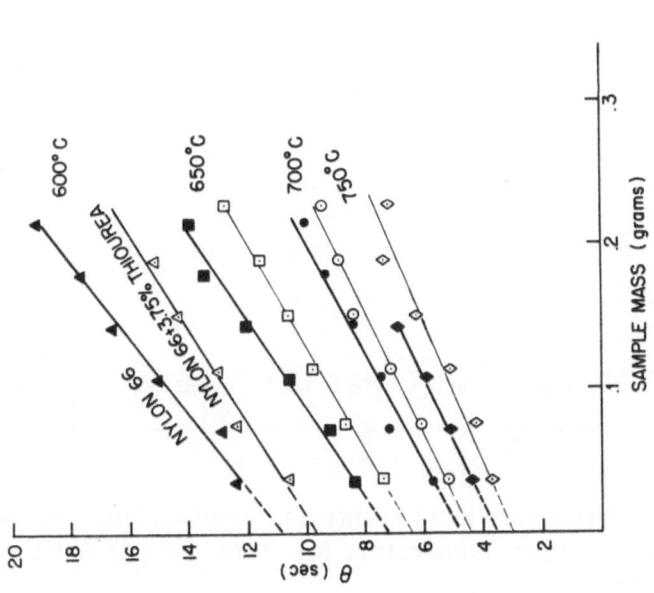

FIGURE 8. Auto-ignition of untreated and flame retardant nylon 66. Filled points = untreated nylon; open points = nylon + 3.75% thiourea.

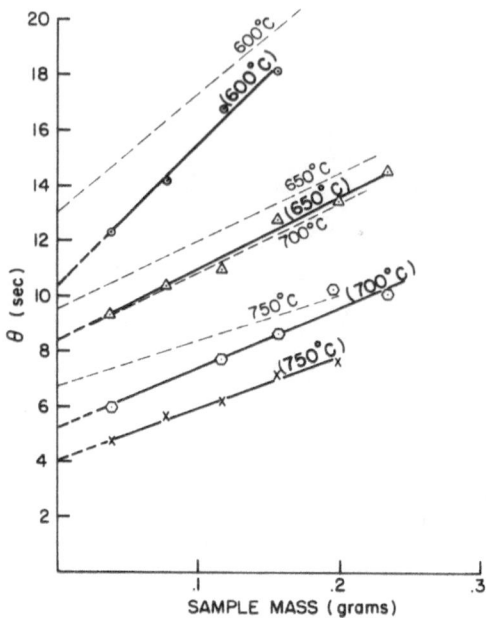

FIGURE 10. Auto-ignition of untreated and flame retardant poly-
(ethylene terephthalate). Solid line = PET + 4.8% DAP; light
broken lines = untreated PET.

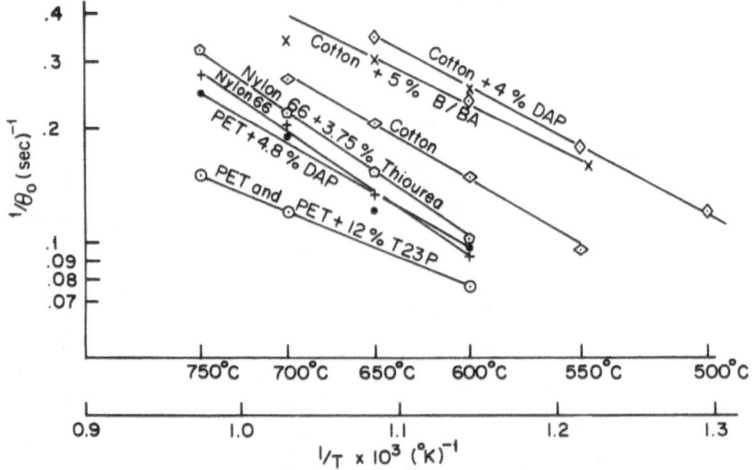

FIGURE 11. Arrhenius plots of intrinsic ignition time (θ_0) as a
function of furnace air temperature for untreated and flame-retar-
dant materials.

Auto-Ignition of Two-Component Blends

The auto-ignition behavior of a 50/50 PET/cotton blend is shown in Figure 12. In comparing these data for the blend with the data for the pure components (Figs. 2 and 10), one notes that for each furnace temperature both the observed and intrinsic ignition times of the blend are less than those of either of the two components.

If both components of the blend can be assumed to have reasonably close decomposition temperatures, during the time interval θ_2 both materials will be decomposing, each at its own rate, and contributing to the combustible gases around the sample. The intrinsic ignition time of the blend represents the combined behavior of two superimposed infinitely thin samples. For this case the total rate of production of combustible gases (or more correctly, the rate of diffusion of the combustible gases into the hot air around the sample) equals the sum of the rates for each component. Since the reciprocal of intrinsic ignition time corresponds to this rate, the intrinsic ignition time of the blend might be derived from the intrinsic ignition times of the two components as follows:

$$(1/\theta_0)_{blend} = (1/\theta_0)_A + (1/\theta_0)_B$$

This represents a typical addition of two rate processes to the same phenomenon; a trivial solution would exist if one of the components is completely inert (i.e., having a high decomposition temperature so that $(\theta_0)_B = \infty$. For this situation, $(1/\theta_0)_B = 0$ and the intrinsic ignition time of the blend would be that of the lower decomposing material. Differences in ignition times at finite masses would be determined by the relative thermal absorptivities of the two components. Such a situation is apparently illustrated by the PET + 12% T23P system in Figure 9.

The equation proposed above represents only a first order approximation to the real physical situation and is based on the assumption that the components do not interact during their thermal pyrolysis (i.e., the pyrolysis behavior of the blend can be derived from the additive sum of the pyrolysis behavior of each of the components.) For the PET/cotton system, studies in

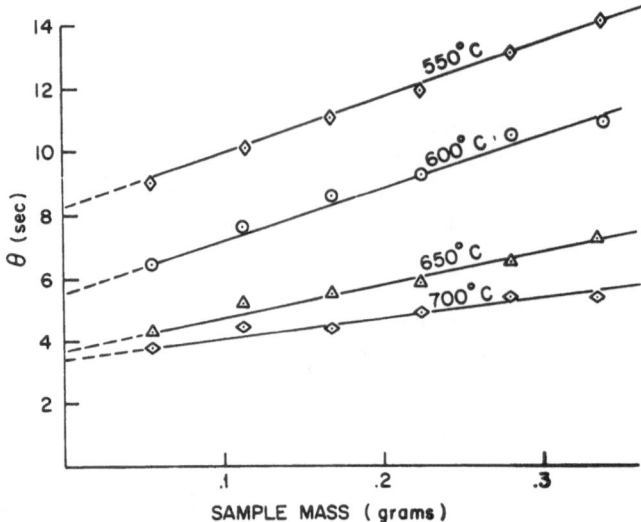

FIGURE 12. Auto-ignition of 50/50 polyester/cotton fabric.

progress[4] have shown that the two components do in fact interact during pyrolysis and that the total production of gas pyroly-
sates from PET/cotton blends is not that which would be predicted
from the additive behavior of the two components. As shown in
Table I, the predicted intrinsic ignition times for the blend are
about 20% less than the observed values. It is very likely that
this represents a manifestation of chemical interaction during
pyrolysis.

TABLE I

AUTO-IGNITION OF 50/50 POLYESTER/COTTON FABRIC

Furnace temp.	(θ_0) cotton	(θ_0) PET	(θ_0) blend (predicted)	(θ_0) blend (observed)
600° C	6.7	13.0	4.4	5.5
650° C	4.9	9.5	3.2	3.7
700° C	3.7	8.3	2.6	3.4

Future work in this area will be concerned with an extension of experiments to other blends in order to establish the validity and limitations of the proposed model.

CONCLUSIONS

Auto-ignition studies of this kind can provide a straight-forward and sensitive means for identifying additives that affect the thermal decomposition behavior of polymers. The technique will also determine how such additives influence the thermal absorptivity of the material before decomposition. Consequently, if an additive has been observed to be a flame retardant (by some reliable burning rate test), but is not found to influence the rate of thermal decomposition, it is likely to be operating only in the gas phase on the flame produced during burning. The work with poly(ester)/cotton mixtures appears to show that chemical interactions during thermal decomposition are detectable through auto-ignition studies.

REFERENCES

1. B. Miller, J. R. Martin, and C. H. Meiser, Jr., J. Appl. Polym. Sci. (in press).

2. A. Alkaidas, R. W. Hess, W. Wulff, and N. Zuber, "Final Report for Government-Industry Research Committee on Fabric Flammability", Office of Flammable Fabrics, National Bureau of Standards, Washington, D. C., December 1971.

3. B. Miller and T. Gorrie, J. Polym. Sci., Part C, No. 36, 3 (1971).

4. B. Miller, J. R. Martin, and C. H. Meiser, Jr., unpublished results.

THERMAL ANALYSIS OF IRRADIATED POLY(VINYL CHLORIDE)

R. Salovey and R. G. Badger

Hooker Chemical Corporation
M. P. O. Box 8
Niagara Falls, New York, 14302

SYNOPSIS

The thermal decomposition and resultant disposal of waste poly(vinyl chloride) (PVC) is facilitated by preliminary exposure to ionizing radiation, such as energetic electrons, in an oxygen atmosphere. The results of isothermal and temperature programmed thermogravimetry, differential thermal analysis and effluent gas analysis in nitrogen and in oxygen indicate that the major effect of irradiation is to render PVC increasingly susceptible to oxidation. The presence of oxygen during heating enhances the decomposition. Crystalline order in PVC is destroyed by irradiation.

INTRODUCTION

The accumulation of waste polymeric materials is a major component of environmental pollution.[1] About one million tons of poly(vinyl chloride) (PVC) are discarded annually. It is suggested that the thermal decomposition of PVC following exposure to ionizing radiation in an oxygen atmosphere would facilitate disposal. Radiation induced oxidation initiates free radical chain dehydrochlorination and the quantitative evolution of hydrogen chloride (HCl).[2] The resultant unsaturated polymer can be used

109

directly, perhaps as a filler, or yield simple hydrocarbons on pyrolysis. The feasibility of this approach is examined by thermal analysis of irradiated PVC.

Irradiation of PVC in vacuo produces free radicals, [3, 4] polyenes, [3, 5] and crosslinks[6] and is accompanied by the evolution of hydrogen chloride. [2] Irradiation in air yields, additionally, chain scission and carbonyl groups. [7] Thermal degradation of PVC produces similar products and has been reviewed. [8, 9, 10]

EXPERIMENTAL DETAILS

The PVC used in these experiments was commercial material prepared by bulk polymerization (Rucon B-20, Hooker Chemical Corp.). The PVC powder was irradiated in air to a series of doses in a beam of 1. 5 MeV electrons from a Van de Graaff generator (State University of New York at Buffalo) equipped with a 14 inch scanner. Samples were exposed to 5. 8, 11. 6, 17. 4, 29. 0 and 52. 2 Mrad. Each pass on a conveyor belt under the beam took 0. 53 minutes and the dose rate was 1.16 Mrad/pass. The dose rate was determined by the bleaching rate of blue cellophane calibrated with the Fricke dosimeter which is the radiation induced oxidation of ferrous ions. [11]

Thermogravimetry (TG) was performed in a vertical tubular oven (~9 mm ID) in atmospheres of oxygen or nitrogen which were renewed 5 times/minute flowing past the samples. Ten mg samples were contained in shallow platinum dishes (~6 mm diameter) and the sample weight monitored with an electrobalance (Cahn RG) and continuously recorded. Thermogravimetry was either isothermal (150° C) or at a heating rate of 10° C/minute. For isothermal thermogravimetry, the furnace with the sample in place was heated from room temperature in about six minutes.

Differential thermal analyses (DTA) were conducted at 10° C/minute on 300 mg samples in contact only with glass. Oxygen or nitrogen gas flowed through the sample so that the atmosphere surrounding the particles was renewed every few seconds.

Effluent off-gases from DTA were led through thin Teflon
tubing into a stirred aqueous scrubber and titrated with 0.1052 N
sodium hydroxide using phenolphthalein indicator.

X-ray diffraction scans of PVC samples were generated
between 2 θ values of 13 and 34° using an X-ray diffractometer
(Norelco) with a copper target X-ray tube (λ = 1.54 Å).

RESULTS AND DISCUSSION

Isothermal thermogravimetry (TG) at 150° C in flowing
oxygen atmosphere is summarized in Figure 1 for control and
irradiated samples.   Data in a flowing nitrogen atmosphere for
the PVC sample irradiated to 52 Mrad are included for comparison.

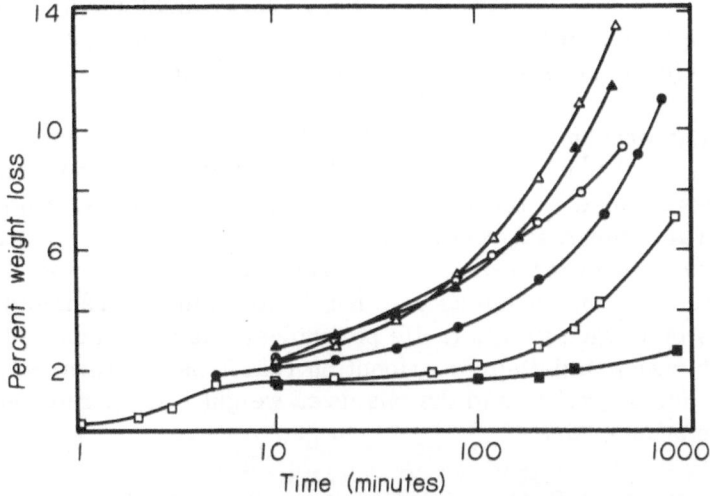

FIGURE 1. Isothermal thermogravimetry of irradiated PVC,
150° C, oxygen atmosphere. ■ control;   □ 5.8 Mrad;   ● 17.4
Mrad; ▲ 29.0 Mrad; △ 52.2 Mrad; ○ 52.2 Mrad, in nitrogen.

Poly(vinyl chloride) undergoes continual weight loss by the evolution of HCl.[12, 13] Initial heating to 150° C produces a weight loss in excess of 1.5% largely independent of irradiation. This may correspond to the volatilization of trapped low molecular weight impurities from synthesis. There is a trend toward an increase in initial weight loss with increasing radiation dose. We infer that the amount of trapped radiolytic HCl is increased. Less initial volatiles were noted at the highest dose (52 Mrad). Perhaps a larger fraction of radiolytic HCl escapes during the irradiation as one prolongs the period of local heating in the radiation field, at the highest dose, leaving less HCl to evolve on initial heating to 150° C. The initial weight loss is independent of atmosphere at 52 Mrad as the initial portion of the weight loss curves coincide in oxygen and nitrogen.

The overall rate of weight loss increases markedly with irradiation dose. This results from a free radical chain dehydrochlorination enhanced by the radiation induced generation of labile chlorine.[13] Crosslinking by irradiation produces tertiary chloride at branch points and radiolytic dehydrochlorination produces allyl chloride. Both structures are labile and are readily susceptible to subsequent thermal decomposition.

In an oxygen atmosphere, the total weight loss is increased compared to that in an inert atmosphere. Peroxides and hydroperoxides formed by the reaction of oxygen with radical fragments from irradiation[7, 14] decompose thermally and induce additional chain dehydrochlorination. The presence of oxygen during TG facilitates decomposition as continual scavenging of transient radicals by oxygen yields labile peroxides. Radical termination by combination and disproportionation reactions are favored in the absence of oxygen and the observed weight loss is reduced in nitrogen atmosphere.

Thermogravimetry of PVC programmed at a heating rate of 10° C/minute in an oxygen atmosphere is illustrated in Figure 2. On heating to temperatures in excess of 100° C, a first stage of weight loss of about 1-2% is observed. Major loss ensues above 150° C corresponding to the evolution of HCl. (The theo-

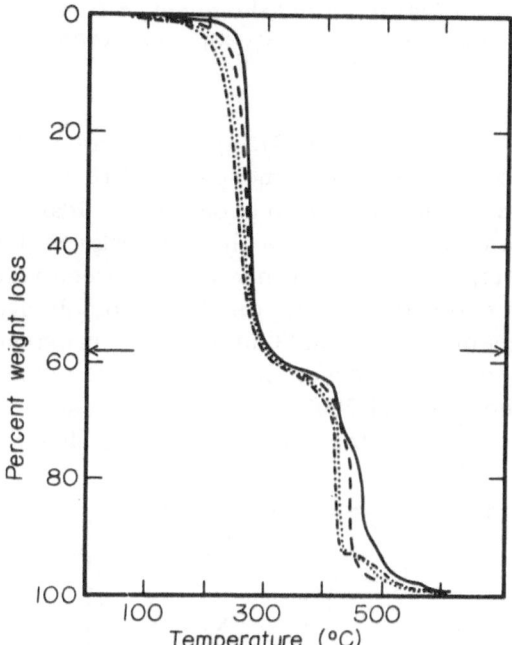

FIGURE 2. Thermogravimetry of irradiated PVC, heating rate, 10°/minute, oxygen atmosphere. ——— control; – – – 5.9 Mrad; ••••• 29.0 Mrad; •–•– 52.2 Mrad.

retical content of HCl is 58% by weight of the PVC molecule and is marked by arrows on the ordinate.)

The temperature dependence of weight loss decreases on further heating until a second stage of decomposition is attained over 400° C. A more complex decomposition pattern is then presented corresponding to main chain fragmentation. [12]

The incipient temperature for rapid weight loss decreases from 200 to ∼160° C on irradiation. The temperature at the onset of rapid decomposition above 70% weight loss is also reduced, for example, from 429 to 413° by irradiation to 5.8 Mrad. The rate of weight loss in this stage of decomposition (70 to 90%) is enhanced

by irradiation. The temperature for the onset of final carboniza-
tion is reduced on irradiation (0 - 52 Mrad) from 476 to 425° C.
The process is complete at ~600° C.

Differential thermal analysis (DTA) of the same samples of
irradiated PVC was pursued in an atmosphere of flowing nitrogen
and at a heating rate of 10° /minute in order to elucidate decomposition patterns. Resultant curves are shown in Figure 3 as a composite plot of the temperature difference (ΔT) between reference
and standard as the ordinate with ΔT displaced for clarity. An
endothermic displacement between 70 and 80° C corresponds to
the glass transition and is observed in all samples. Thermal
activity is next detected about 200° C and may correspond to the
final melting of PVC crystallites. This feature is absent follow-

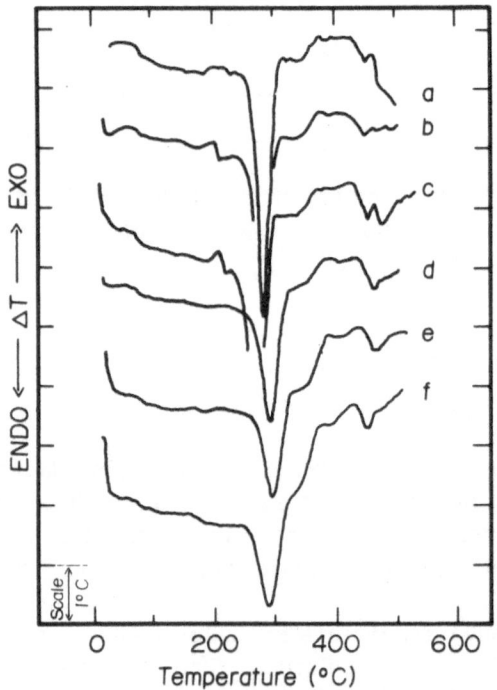

FIGURE 3. Differential thermal analysis of irradiated PVC, heating rate, 10°/minute, nitrogen atmosphere. a, 0 Mrad; b, 5.8
Mrad; c, 11.6 Mrad; d, 17.4 Mrad; e, 29.0 Mrad; f, 52.2 Mrad.

ing irradiation at doses ≥ 17.4 Mrad. From DTA we would infer that crystalline order is destroyed by irradiation. X-ray determinations were performed on duplicate PVC samples and low levels of crystallinity are estimated. [15, 16] Control and PVC samples irradiated to 5.8 Mrad have well-defined crystalline diffraction peaks at 2 θ values of 17, 19 and 25°, indicating crystallinities of 20 and 15%, respectively. Following 52 Mrad irradiation no crystalline diffraction peaks are discernible.

Simultaneous effluent gas analysis (EGA) in nitrogen atmosphere is reported in Figure 4. These data are consistent with DTA of PVC in nitrogen in which the major feature is a large endotherm between 250 and 300° C assigned to dehydrochlorination. Increasing irradiation dose facilitates the evolution of HCl at low temperatures. Consequently, residual HCl is more tenaciously bound.

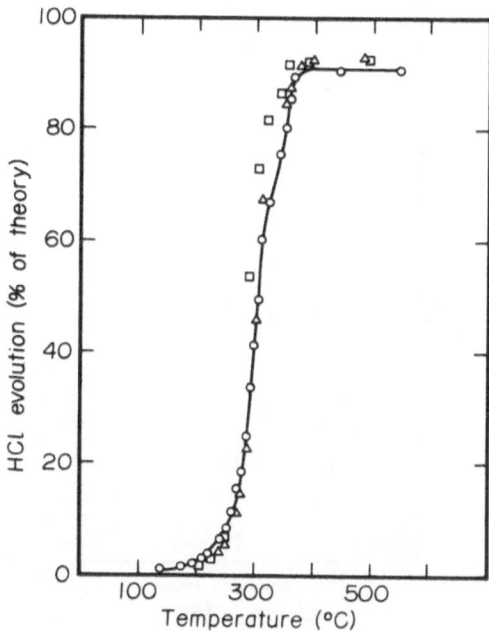

FIGURE 4. Effluent gas analysis of irradiated PVC, heating rate, 10°/minute, nitrogen atmosphere. □ 5.8 Mrad; △ 17.4 Mrad; ○ 29.0 Mrad.

Analogous to Figure 3, a composite plot of DTA in flowing oxygen is shown in Figure 5. Again, the curves are displaced for clarity. Thermometric patterns are considerably affected by the presence of oxygen during heating. An endothermic displacement associated with the glass transition at ~75° C in unirradiated PVC becomes less prominent with increasing dose. Further heating reveals a peak at 200° C associated with crystalline melting. This is not observed at doses in excess of 17.4 Mrad. Crystalline melting is observed in the 17.4 Mrad sample heated in oxygen but not in nitrogen. This may be due to greater chain scission occurring on heating in oxygen. The scission induced crystallization of high molecular weight poly(tetrafluoroethylene) has been reported. [17]

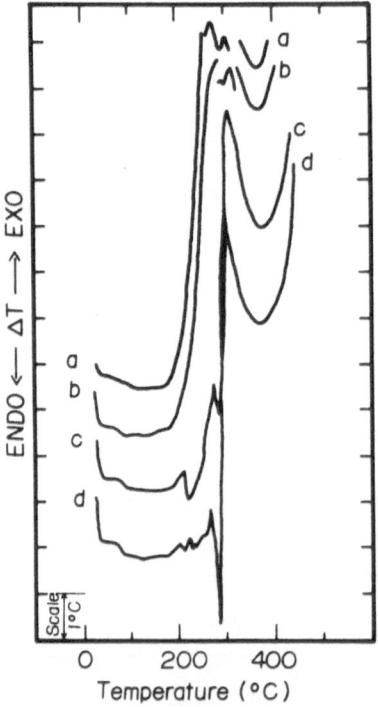

FIGURE 5. Differential thermal analysis of irradiated PVC, heating rate, 10°/minute, oxygen atmosphere. a, 52.2 Mrad; b, 29.0 Mrad; c, 5.8 Mrad; d, 0 Mrad.

A huge exotherm is observed in Figure 5 between 200 and 300° C and is absent in Figure 3. This occurs at lower temperatures with increased dose. This is associated with oxidation and may indicate the decomposition of peroxides formed on irradiation and on heating in oxygen. It is likely that at all irradiation doses endothermic dehydrochlorination is superposed on oxidation. At doses in excess of 17.4 Mrad, this endotherm is largely obscured. We suggest that the effect of irradiation is to render PVC increasingly susceptible to oxidation.

Effluent gas analysis of PVC in flowing oxygen is shown in Figure 6. Again dehydrochlorination observed between 250 and 300° C coincides with a calorimetric endotherm. A comparison of EGA in oxygen and nitrogen atmospheres indicates that the evolution of HCl follows a similar temperature pattern with dehydrochlorination occurring about 15° C lower in oxygen.

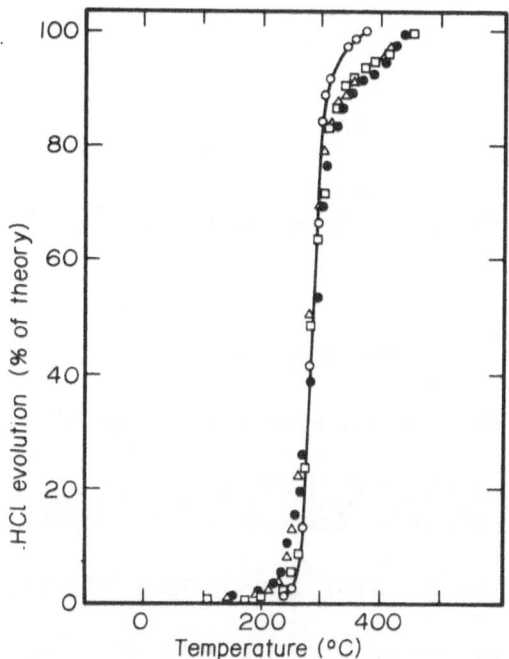

FIGURE 6. Effluent gas analysis of irradiated PVC, heating rate, 10°/minute, oxygen atmosphere. ○ control; □ 5.8 Mrad; △ 17.4 Mrad; ● 29.0 Mrad.

CONCLUSIONS

The isothermal rate of decomposition of pre irradiated PVC on heating at 150°C increases with irradiation dose and is enhanced in oxygen atmosphere. The increased thermal instability is consistent with the radiation induced generation of labile chlorides and, in oxygen, labile peroxides. The two major stages of thermal decay observed on programmed heating of irradiated PVC occur at reduced temperatures after irradiation. From X-ray and DTA results, it is inferred that crystalline order in PVC is lost on irradiation. Dehydrochlorination of PVC is endothermic. In an oxygen atmosphere, the DTA behavior of preirradiated PVC is dominated by oxidation. The major effect of irradiation is to render PVC increasingly susceptible to oxidation. The radiation damage appears to be nonlinear with maximal effect at lower doses.

ACKNOWLEDGMENT

We are grateful to I. M. Zsolnay for the X-ray diffraction measurements and to A. L. Martyn for his help in thermal analysis.

REFERENCES

1.   T. C. Purcell, Polymer Preprints, 12, no. 2, 81 (1971).

2.   R. Salovey, in The Radiation Chemistry of Macromolecules, vol. 2, Ed., M. Dole, Academic Press, New York, 1973.

3.   G. J. Atchinson, J. Appl. Polym. Sci., 7, 1471 (1963).

4.   B. R. Loy, J. Polym. Sci., 50, 245 (1961).

5.   R. Salovey, J. P. Luongo, and W. A. Yager, Macromolecules, 2, 198 (1969).

6.   M. A. Crook and F. S. Walker, Nature, 198, 1163 (1963).

7.   R. Salovey and R. C. Gebauer, J. Polym. Sci., Part A-1, 10, 1533 (1972).

8.   M. Asahina and M. Onozuka, J. Polym. Sci., Part A, 2, 3505, 3515 (1964); M. Onozuka and M. Asahina, J. Macro-mol. Sci., 2, 235 (1969).

9.   D. Braun, Pure Appl. Chem., 26, 173 (1971).

10. W. C. Geddes, Rubber Chem. Tech., 40, 177 (1967).

11. A. J. Swallow, Radiation Chemistry of Organic Compounds, Pergamon, New York, 1960, p. 42.

12. R. R. Stromberg, S. Straus, and B. G. Achhammer, J. Polym. Sci., 35, 355 (1959).

13. R. Salovey and H. E. Bair, J. Appl. Polym. Sci., 14, 713 (1970).

14. R. Salovey, R. V. Albarino and J. P. Luongo, Macromole-cules, 3, 314 (1970).

15. J. A. Manson and S. A. Iobst, personal communication.

16. R. J. d'Amato and S. Strella, Appl. Polym. Symposia, 8, 275 (1969).

17. W. R. Licht and D. E. Kline, J. Polym. Sci., Part A-2, 4, 313 (1966).

RECYCLING POLY(TETRAFLUOROETHYLENE)

Barry Arkles

Liquid Nitrogen Processing Corporation
412 King Street
Malvern, Pennsylvania, 19355

Poly(tetrafluoroethylene) (PTFE) is unique among polymers in its ecological position. Although there has been considerable discussion on recycling waste polymers such as poly(ethylene)
(PE), poly(styrene) (PS), and poly(vinyl chloride) (PVC), only
poly(tetrafluoroethylene) has been recycled in quantities significant compared to production. Over fifty times more PE is consumed in the form of trash bags for waste disposal than PTFE
manufactured. Further, the growth rate of PE, PVC, and PS is
continuing at 5-10% per year while PTFE growth has plateaued.
On a tonnage basis, then, the recycling of PTFE appears anomalous.
The discrepancy disappears when the economic reward of reprocessing a \$2.75/lb material is contrasted with that of a \$0.14/lb
material. The high scrap value of PTFE provides an incentive
for both fabricators and end-users to accumulate waste in reasonably concentrated bulk. Further, freight cost to reprocessing
installations is not a significant portion of the polymer cost. Due
to the economic incentive a wide variety of methods for up-grading
waste PTFE to reprocessable grades have been evaluated. The
techniques employed may be transferable to other polymers of
greater ecological impact.

The fate of PTFE under normal conditions of weathering
varies from other polymeric materials because fewer paths of
degradation are open to it. Thermoplastic resins typically

121

contend with the oxidative, hydrolytically-catalyzed oxidative,
and chemically-induced degradative processes that characterize
disintegration of metals that predate them in many applications.
Additionally, they are susceptible to thermal, bacterial and,
most importantly, photochemical degradation processes. Thermal
and photochemical degradative processes result in polymer chain
scission and cross-linking, but the thermal process proceeds
through lower energy vibrationally excited states while ultraviolet radiation induces degradation through higher energy elec-
tronically excited states. Poly(ethylene) is exemplary: ultraviolet irradiation of PE results in free radical formation and
subsequent oxidation forms hydroperoxides. Carbonyl formation and chain scission generally follow, and mechanical and
erosive forces serve to disintegrate the physical structure when
photo-oxidative degradation has progressed sufficiently. Finally
bacteriological degradation of the low molecular weight fragments proceeds to carbon dioxide and water.

Normal pathways of degradation are disallowed for poly-
(tetrafluoroethylene). The atmosphere blocks radiation with sufficient energy to excite the carbon fluorine bond. Carbon oxygen
bond formation in preference to carbon-fluorine is unfavorable
thermodynamically due to the latter's greater stability. PTFE
is not wet by water and consequently hydrolysis and hydrolytically-
catalyzed degradation do not occur. Bacteriological degradation,
like most of the other processes, requires abstractable hydrogen,
so PTFE does not behave as a substrate for enzymatic reactions.
In the event that any of these unlikely events occur, consideration must still be given to the fact that the molecular weight of
PTFE resins is generally four to six million, whereas other
commercial polymers have molecular weights in the range of
6,000 to 400,000. Thus, PTFE must undergo considerably
greater degradation to result in polymer fragments the same
size as other commercial polymers. Only two pathways of degradation are open to PTFE. The resin is extremely susceptible to
high energy radiative degradation. A dosage of 5 megarads will
degrade PTFE integrity. Although one-hundredth the dosage
required for more typical polymers, this is still considerably
greater than the exposure seen under normal weathering. The

only practical pathway of PTFE breakdown is through erosive forces. Little shear energy is required to initiate adhesive wear of PTFE. The wear rate of PTFE against steel, for example, is 10 to 100 times greater than the typical thermoplastics (wear factor $K = 2 \times 10^{-7} \text{ in}^3 \text{ min}/(\text{ft})(\text{lb})(\text{hr})$). Thus the normal fate of PTFE in the environment is analogous to that of inorganic debris.

The world market for PTFE is variously estimated between 23 and 26 million pounds per year. In the United States more than half of the production is dedicated to dispersion and coagulated dispersion grades. The balance is comprised of granular resins including presintered and composite grades. Dispersion and composite grades are showing the greatest growth, but during the last three years overall growth has dwindled to less than 5% for PTFE resins. The limit in demand for the resin is due to its failure to lend itself to automatic fabrication, its high cost, and poor dimensional stability. Compared to the seven billion pound per year growing market for poly(ethylene), poly(tetrafluoro-ethylene) is of trifling environmental concern. Further, it is generally utilized in engineering rather than packaging applications and does not contribute to litter.

At present 1.0 - 1.4 million pounds per year of granular PTFE is recycled. The principal part of this material is upgraded at installations distinct from that of the fabrication. Hand and mechanical sorting account for most of the upgrading. Material is reduced to a 40 - 80 mesh particle size and returned to fabricators. Processing differs from virgin material in that the simultaneous application of pressure and heat is required where virgin material is generally compressed and sintered in separate steps. The most common techniques for fabricating reprocessed PTFE (repro) are ram extrusion, coining, and pressure sintering. Table I compares the mechanical properties of virgin and repro grades of PTFE. In general, physical properties are lowered 50% while electrical properties are maintained. The reduction of physical properties does not place serious restrictions on end use of repro since PTFE parts do not often perform in load-carrying applications. Typical applications are bearings, valve seats, and gaskets.

TABLE I

POLY(TETRAFLUOROETHYLENE): VIRGIN vs.
REPROCESSED RESIN PROPERTIES

Property	Test method	Virgin	#1 repro
Specific gravity	ASTM D1457-62T	2.14 - 2.20	2.15 - 2.20
Tensile strength (P.S.I.)	ASTM D1457-62T	3500 - 6000	1600 - 2400
Elongation (%)	ASTM D1457-62T	150 - 450	100 - 200
Dielectric strength (volts/mil)	ASTM D149-64	1500	1200
Surface resistivity (ohms)	ASTM D257-61	10^{16}	10^{16}
Volume resistivity (ohm-cm)	ASTM D257-61	10^{17}	10^{17}
Dielectric constant	ASTM D150-64T	2.10	2.26
Dissipation factor	ASTM D150-64T	0.0003	0.0004

Feedstock for repro production has not only been limited to granular resins but to specific types of granular based materials. The usable forms have generally consisted of hard stock solid pieces of fabricated material, e.g., waste rod stock or used shuttle pads as used in the textile industry, and machine turnings. Reprocessors pay a higher premium for hard stock. The material is further graded as to external and internal cleanliness, surface contamination, and segregatable internal contamination. A progressively lower price is paid in each case.

The amount of reusable PTFE is considerably less than the total amount available. It is estimated that the total available volume of waste PTFE is between four million and five million pounds per year. Aside from resins with non-segregatable

contamination, oriented paste, composite blends and forms,
material with more than one heat history and pigmented material
have not been utilized. A variety of new techniques have been
investigated and several are in initial stages of commercial
development.

In order to design a process for producing reprocessable
grades of poly(tetrafluoroethylene), the contamination must be
removed and the molecular weight and crystallinity must be amenable to normal processing techniques. Table II summarizes the
contamination most often found with PTFE. A variety of polymers
with high thermal and chemical resistance are found among PTFE
scrap since these materials often are not only used in the same
applications, but fabricated in the same shops. While some contamination may be unexplainable, as exemplified by wads of paper,
in composites high levels of carbon, bronze and glass may be
uniformly distributed throughout the resin. Pigmentation also
provides special problems since these are usually highly stable
inorganic compounds. Common pigments for fluorocarbon resins
are cadmium sulfide (yellow); cadmium selenide (red); chromium
and cobalt compounds (green); chromium, cobalt and aluminum
compounds (blue); cadmium and mercury sulfides (orange); and
manganese, cobalt and aluminum compounds (violet). Because
of the wide range of contamination found in PTFE it becomes
easier to conceptualize a process that discriminates PTFE from
contamination, rather than a process which discriminates contamination from PTFE. It therefore becomes very important to
examine the properties of PTFE resins and determine which ones
can be exploited alone or in combination to separate PTFE from
contamination.

Salient characteristics of PTFE include its extreme chemical resistance. The resin reacts only with molten alkali metals,
fluorine gas, and fluorine interhalogen compounds. It possesses
greater thermal stability than most organic compounds due to the
fact that the carbon-fluorine bond is about 6 kcal/mole more
stable than the carbon-hydrogen bond, as well as the fact
that the larger size of the fluorine causes a sterically closed
structure in PTFE. The high fluorine content gives PTFE a
specific gravity of 2.14 - 2.19, considerably higher than other

TABLE II

CONTAMINATION FOUND IN SCRAP
POLY(TETRAFLUOROETHYLENE)

General category	Specific breakdown
Resins	Polyamide Chlorotrifluoroethylene High density poly(ethylene) Poly(methylene) oxides

Metals	Iron
	   magnetic
	   non-magnetic
	Aluminum
	Brass

Organic	Cellulose wood fibers, paper, etc. Oil Carbon
Inorganics	Alumina Silica Molybdenum disulfide Iron oxide Pigments

polymers. It possesses an extremely low critical surface tension,
18. 5 dynes/in, and thus is not wet by most solvents. When ground
to a particle size less than 150 microns, PTFE will float on water
even though it is twice as dense because its resistance to wetting
is so great. PTFE is an insulator with high capacitance. It is
not magnetic. An important although seemingly trivial aspect is
its optical transmission through the visible and ultraviolet spectrum. It absorbs strongly in the lower frequency infrared.

Chemical treatments of two distinct types have been employed
for the removal of loose surface contamination and internal contamination. While the removal of surface contamination is current
art, implementation of a variety of systems for removal of internal contamination is a recent development. Removal of surface
contamination is accomplished either through use of aqueous or
non-aqueous washing systems. In an aqueous system the scrap
is essentially laundered in a water solution of strong soap (e. g.,
metasilicate) and a non-ionic surfactant (e. g., Igepal). Solvent
systems may be either chlorinated or non-chlorinated hydrocarbons.
They are particularly applicable for machine oil contaminated
material. Either extraction or vapor degreasing systems are
utilized. Removal of surface contamination does not amplify feedstock for reprocessing. The more difficult problem is to utilize
the large amount of scrap PTFE which contains internal contamination.

Internal contamination may be introduced deliberately as,
for example, in composite or pigmented resins or inadvertantly
as, for example, "spot" inorganic or "bloom" organic contamination. In each case a method must be provided for removal of contamination from the resin matrix. Examples of the systems used
include nitric acid, fuming nitric acid, nitric acid - perchloric
acid (liquid fire), dichromate - acid, and ozone. These sytems
exploit the fact that small molecules can permeate the PTFE and
react with contaminants, converting them to ionic species or
gases. Each of the systems has limitations. While nitric acid
can be used to remove metals, it is slow to oxidize organics that
have withstood sintering. It has little effect on inorganic pigments.

Fuming nitric acid[1] has been utilized for the removal of pigmentation in finished as well as scrap PTFE products. An example is the removal of color coding from PTFE insulated wire. Coiled wire is placed in a stainless steel tank and covered with the acid for five minutes. The bulk of the acid is removed and a small amount of water is added liberating N_2O_4, N_2O_3, NO, and N_2O. This completes oxidation and nitration of the pigment. The coil is thoroughly rinsed with water and can be used as white code-free wire. Combinations of nitric and perchloric acids[2] have been utilized in the "liquid fire" treatment of scrap. The nitric acid is utilized as a mildly oxidizing diluent while the mixture is brought to a temperature for perchloric acid activation. The perchloric acid oxidizes most organic material to carbon dioxide and converts most inorganic material to soluble perchlorates. Systems which concentrate on removing organic contamination include ozone and dichromate - mineral acid mixtures. These are less versatile systems and often convert inorganic contamination to highly colored compounds.

It is clear from this brief discussion that there are severe limitations to chemical treatment methods. The methods are hazardous. Aside from exposure to acidic and toxic materials, an explosion hazard exists in most of the methods if an easily oxidized material is introduced with the scrap. If large quantities of foreign material are present, reaction time may be too long to be economical. Inorganic composites and non-granular PTFE resins cannot be upgraded by this method.

One approach to expanding the utility of the chemical methods is to provide a feedstock which is free of bulk contamination. Machine turnings of PTFE, for example, are frequently entangled with machine turnings of other materials, soiled with abrasive grindings, and contaminated with bits of tool metal. The sedimentation properties of PTFE offer a convenient method for the removal of bulk contamination. One such method is based on air classification. Air classification has been proposed by the Stanford Research Institute[3] for the sorting of municipal refuse. A zig-zag configuration air classifier is depicted in Figure 1. If a particle is introduced into an upward directed air-stream, it will either settle or travel upward according to its density and drag coefficient.

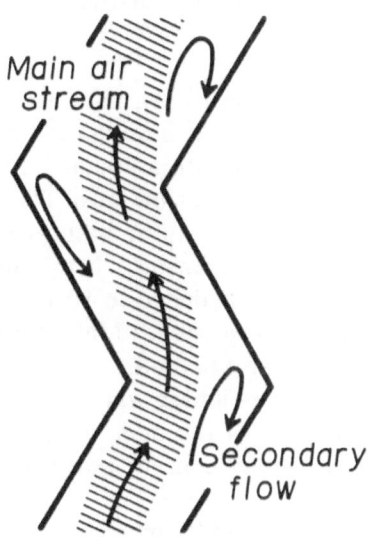

FIGURE 1. Zig-zag configuration air-classifier.

$$Vp - Va \propto - \left[\frac{dp}{Co} \frac{\rho_p}{\rho_a} \right]^{1/2}$$

where   Vp = velocity of particle   Co = drag coefficient
             Va = velocity of air-stream ρ_p = particle density
             dp = particle diameter ρ_a = air density

Operationally, chopped material is introduced one-third from the
top of the classifier. Lighter material is blown out the top while
denser material cascades out the bottom. Presumably, by varying
air velocity in two sequential operations, chopped PTFE could
first be carried over as the lighter material to a second classifier
where it would sediment. First heavy, then light, contamination
would be removed. The refinement appears analogous to distillation, but in practice it is more difficult. Materials like aluminum (2.70 g/cc) and glass (2.54 g/cc) have densities close to
PTFE (2.15 g/cc). Particle size and configuration vary the drag
coefficient, making it impossible to maintain uniform operating

conditions. While this method may be promising for separating municipal refuse, it has only limited utility in upgrading scrap PTFE.

A second method for utilizing the sedimentation properties of PTFE is underwater milling. If a feedstock contains cellulose, poly(ethylene), or any material with a specific gravity less than 1.00, they can be removed by flotation from PTFE. A single pass through an underwater mill into a water bath provides sufficient energy to separate most contamination. If the underwater milling is continued until the PTFE is reduced to 80 mesh or smaller, it floats due to its resistance to wetting. Materials with densities greater than 1.00 sink. As an example, a 25% glass fiber composite was milled three passes under water with separation of floating material each time. At the end, with an average particle size of 80 –100 mesh, the glass content was reduced to 1.9%.

Metals are easily removed by most of the methods described above. Additionally, two other methods are employed. The utilization of magnets for ferromagnetic materials is widespread. Non-magnetic metals may be removed from a moving product stream by inductively controlled metal separation units. Metals passing through a high frequency electromagnetic field produce interference which is amplified and used to trip a switch which momentarily deflects the product stream. These systems are ordinarily incorporated into packaging lines in order to safeguard against tramp metal introduced during processing.

The use of the above processes is expected to nearly double the amount of recycled PTFE by 1978. This is a significant increase, but demand for repro grades of PTFE will continue to exceed availability. A method for matching demand must provide for utilization of all composite materials, including the particularly difficult carbon/graphite blends, highly crystalline resin including scrap reprocessed material, and dispersion grades.

A new method which is demonstrating the greatest promise for recycling PTFE materials is a two stage treatment of scrap with super-heated steam. In the first exposure, organic contaminants are removed. On the second exposure, PTFE is depolymerized to tetrafluoroethylene monomer. The monomer may be purified and repolymerized by existing techniques to virgin grade polymer. Cost of monomer from the depolymerization of scrap PTFE is lower than that from difluorochloromethane, the common commercial intermediate.

An apparatus for super-heated steam treatment of scrap PTFE is shown in Figure 2. The first stage of treatment is depicted. Scrap enters a 625°F steam fluidized bed through an air lock. The bed is at a slight incline. While moving over it, materials with less thermal stability than PTFE are decomposed to a variety of volatile products. The steam and volatiles are separated from the resin at the end of the bed. Specific routes are followed by the various resin contaminants. Light machine oil is removed rapidly by steam distillation. For higher molecular weight oils and poly(ethylene) a prior step of molecular weight reduction by chain scission is required. [4, 5]

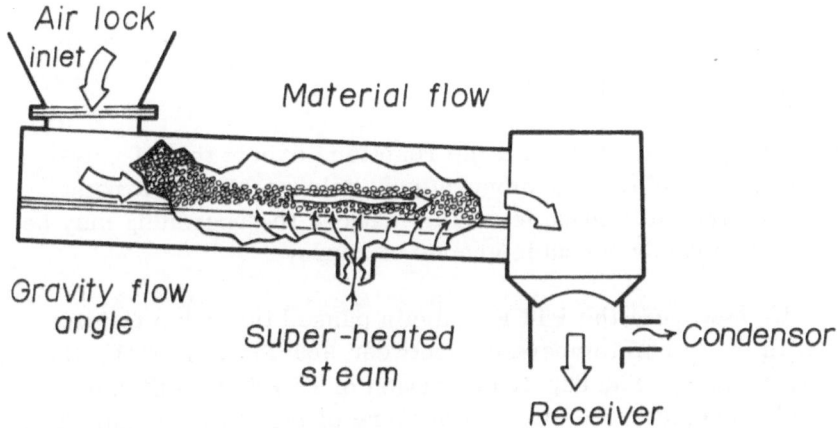

FIGURE 2. Super-heated steam fluidized bed, inclined.

$$\underset{\substack{|\ |\ |\ |\ |\ | \\ H\ H\ H\ H\ H\ H}}{\overset{\substack{H\ H\ H\ H\ H\ H \\ |\ |\ |\ |\ |\ |}}{\sim C-C-C-C-C-C\sim}} \longrightarrow \underset{\substack{|\ |\ |\ | \\ H\ H\ H\ H}}{\overset{\substack{H\ H\quad H \\ |\ |\quad |}}{\sim C-C-C=C}} + \underset{\substack{|\ | \\ H\ H}}{\overset{\substack{H\ H \\ |\ |}}{H-C-C-}}$$

Poly(chlorotrifluoroethylenes) are reduced to low molecular weight waxes and oils which are volatized. [6]

$$\underset{\substack{|\ |\ |\ |\ |\ | \\ F\ F\ F\ F\ F\ F}}{\overset{\substack{Cl\ F\ Cl\ F\ Cl\ F \\ |\ |\ |\ |\ |\ |}}{\sim C-C-C-C-C-C\sim}} \longrightarrow \underset{\substack{|\ |\ | \\ F\ F\ F}}{\overset{\substack{Cl\ F\ Cl \\ |\ |\ |}}{\sim C-C-C-Cl}} + \underset{\substack{|\ |\ | \\ F\ F\ F}}{\overset{\substack{F\quad F \\ |\quad |}}{C=C-C\sim}}$$

Polyamides are hydrolyzed to monomers or comonomers.

$$\underset{HOH}{\underset{\substack{|\ |\quad\ \ |\ |\ | \\ H\ H\quad H\ H\ H}}{\overset{\substack{H\ H\ O\quad H\ H \\ |\ |\ \|\quad |\ |}}{\sim C-C-C-N-C-C\sim}}} \longrightarrow \longrightarrow \underset{\substack{|\ | \\ H\ H}}{\overset{\substack{H\ H\ O \\ |\ |\ \|}}{\sim C-C-C-OH}} + \underset{\substack{|\ |\ | \\ H\ H\ H}}{\overset{\substack{H\ H\ H \\ |\ |\ |}}{N-C-C\sim}}$$

Poly(methylene) oxides are depolymerized by unzipping at chain ends to formaldehyde.

$$\underset{\substack{|\quad\ |\quad\ | \\ H\quad H\quad H}}{\overset{\substack{H\quad H\quad H \\ |\quad\ |\quad\ |}}{\sim C-O-C-O-C-\ddot{O}-H}} \longrightarrow \underset{\substack{|\quad\ | \\ H\quad H}}{\overset{\substack{H\quad H \\ |\quad\ |}}{\sim C-O-C-O-H}} + \overset{\substack{O \\ \|}}{H-C-H}$$

The activation energy for all the reactions except that of poly-
(ethylene) is achieved at temperatures below PTFE melt point.
For scrap free of inorganic contamination, the upgrading may be
sufficient to produce usable grades of repro.

If, however, the PTFE is again passed through a fluidized
bed with the steam temperature between 950° F and 1250° F, it
depolymerizes. The depolymerization of PTFE is well known,
but under ordinary conditions a mixture of many compounds including hexafluoropropene, octafluorocyclobutane, perfluoroisobutene
and other polymeric products having low melting points are
evolved. [7] These products usually contain considerably less than

50% tetrafluoroethylene, and consequently the mixture is of little
commercial value since the individual compounds may be separated
only by elaborate fractionation. The yield has been enhanced by
conducting the pyrolysis under vacuum. Practical yields of 85%
have been reported at pressures below 150 mm Hg. [8, 9] This
improvement could not, however, be commercialized since the
use of vacuum entails a batch process with external heating. Difficulties arise for several reasons: materials capable of with-
standing the high temperature and corrosive pyrolysis conditions
generally have poor heat transfer characteristics; the use of
vacuum eliminates any atmospheric heat convection over the polymer and thus further curtails the supply of energy necessary for
pyrolysis; and poly(tetrafluoroethylene) itself has very poor heat
transfer characteristics. The net result is that the maximum
diameter of the pyrolysis vessel is limited to about 10 inches.
Super-heated steam provides a more efficient path to the result
effected by vacuum. [10] The mechanism for the decomposition
of poly(tetrafluoroethylene)

suggests that by-products arise from the reaction of tetrafluoroethylene itself. In fact, the monomer may undergo 2 + 2 cyclo-
addition or dissociation to difluorocarbene which can react with
tetrafluoroethylene. These reactions are shown in Figure 3. If
PTFE is depolymerized by super-heated steam on a fluidized bed,
tetrafluoroethylene is diluted as formed and contact time is di-
minished. If the mole ratio of steam to depolymerization products
is four to one, yields of monomer equivalent to those achieved
under vacuum are achieved. Increasing the proportion of steam
increases the yield of monomer. A comparison of depolymerization products is given in Table III.

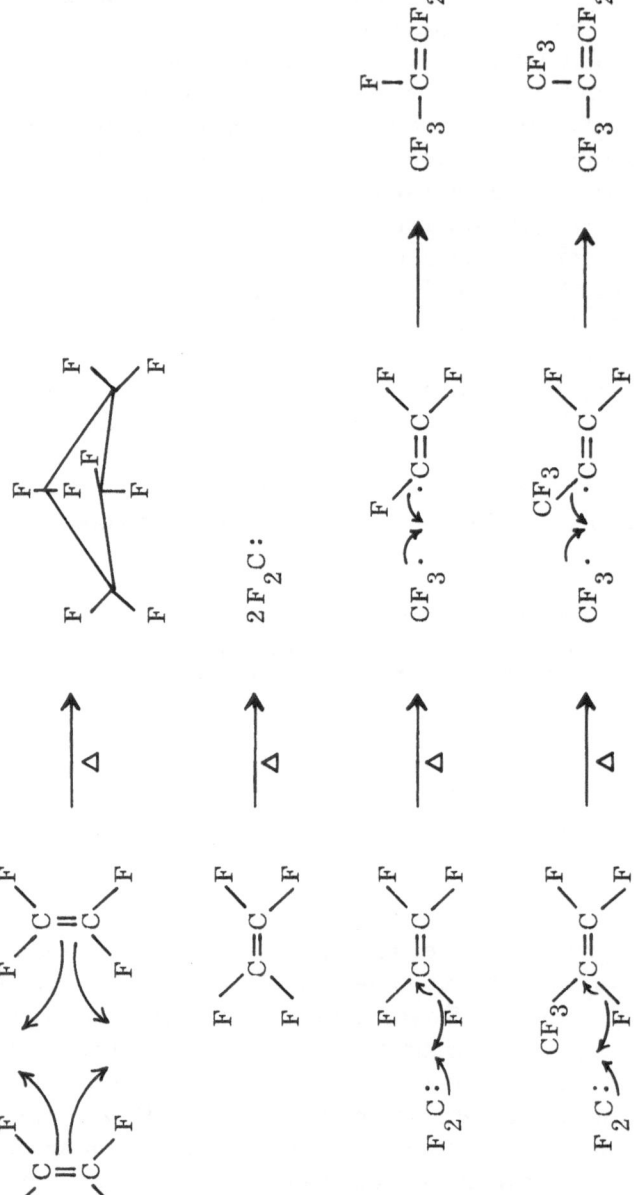

FIGURE 3. By-products generated by tetrafluoroethylene during poly(tetrafluoroethylene) depolymerization.

TABLE III

DEPOLYMERIZATION PRODUCTS OF POLY(TETRAFLUOROETHYLENE)

Depolymer- ization products	Desig- nation	Atmos- pheric	Vacuum <150 mm pressure	Super- heated steam 5:1	Super- heated steam 60:1
C_2F_4	TFE	26.5	84.1	86.4	98.0
C_3F_6	HFP	15	15.3	9.3	1.3
C_4F_8	PFIB	33	—	3.5	0.7
Other	—	25	—	0.8	—

Paste resins, sintered composites, pigmented material, degraded resins, etc. have all been depolymerized in high yield to tetrafluoroethylene by super-heated steam. Monomer has been purified and used for the production of virgin polymer. In contrast to the methods discussed previously, all forms of scrap may be converted in a rapid low-cost operation to feedstock for new polymer preparation.

One additional method of possible future impact was evaluated. If consideration is given to the fact that most of the contamination found in PTFE differs from it by not being white, optical methods are of potential interest. A product stream spread thinly on a conveyor belt should expose most "black spots". All that remains is to remove them while removing as little as possible of the good material. A Skinnerian pigeon could see and peck a spot fairly efficiently, but the author's prejudice directed his attention to optical sorting systems. In order to remove a spot with minimum waste, it must be pinpointed accurately. Accurate pinpointing suggested laser optics. Light is absorbed by dark objects and reflected by light ones. It seen followed that if sufficient energy is absorbed by a particle, it would be vaporized. Thus, a high

power laser could presumably sweep a moving products stream
and vaproize any non-white materials. Two types of high power
lasers are available commercially, carbon dioxide and YAG
(yttrium-aluminum-garnet). The carbon dioxide laser unfortunately operates in the infrared region where PTFE absorbs.
YAG lasers are available only at 100 watt maximum power. Although it has sufficient power to vaporize metal, the duration of
exposure (0.01-0.1 sec) is too long to be practical. The machine
carbonizes nylon, but does not vaporize the carbon. In two to
three years, with the advent of 1000 watt lasers, the method may
be fruitfully reevaluated.

An attempt has been made to demonstrate the various ways
in which the properties of a polymer can be exploited for upgrading in order to recycle it. The specific material reviewed, PTFE,
while resistant to weathering is not an important ecological factor.
Commercialization of recycling technology has developed as a
consequence of the demand for low cost grades of PTFE resin.
Of the several technique reviewed, super-heated steam depolymerization followed by repolymerization promises the greatest utility.

REFERENCES

1. A. J. Perreault, U. S. Patent 3,440,235, April 22, 1969.

2. G. F. Smith, The Wet Chemical Oxidation of Organic Com-
    pounds Employing Perchloric Acid, G. F. Smith Co.,
    Columbus (1965).

3. C. G. Gunnerson and C. E. Wooldridge, "Air Classification
    of Municipal Refuse", Stanford Research Institute Research
    Brief, Irvine, June 11, 1971.

4. S. L. Madorsky, Thermal Degradation of Organic Polymers,
    Wiley, New York (1964).

5. L. A. Wall, Fire Research Abstracts and Reviews, 13,
    204 (1971).

6. German Patent 1,003,700.

7. A. F. Benning, U. S. Patent 2,394,581, February 12, 1946.

8.   E. E. Lewis, U. S. Patent 2,406,153, August 20, 1946

9.   E. E. Lewis and M. A. Naylor, J. Amer. Chem. Soc., <u>69</u> 1968 (1947).

10. B. C. Arkles, U. S. Patent Application, Ser. No. 111,963.

AN INDUSTRY VIEW OF PLASTICS IN THE ENVIRONMENT

George W. Ingle

Monsanto Company
1101 17th Street, N.W.
Washington, D. C., 20036

INTRODUCTION

The title of this paper is meant to emphasize that there is no single monolithic industry viewpoint. Perhaps this diversity is most evident in considering plastics with accelerated degradation. We will later refer to the merits of plastics' characteristic resistance to degradation in landfill operations. Many plastics formulations have already been announced which show some acceleration of degradation on exposure to terrestrial UV, IR, moisture, some combination thereof, and also to certain bacteria and fungi in soil. Claims have been made for plans to develop commercially one or more varieties. One has appeared in this country, with undocumented claims for so-called Food and Drug Administration "approval" since it does contact food. If FDA's recent proposal requiring environmental impact statements (for food additives, including food packaging) becomes regulation, with review by the Environmental Protection Agency, there will clearly be needed explicit analysis of the real significance to the environment.

Some of these recent developments may reflect a "chemical tour-de-force" in reversing past decades of stabilization to create materials with some degree of controlled "self-destruction". But if we take a broader view, to "internalize the externalities",

139

we find some doubts in the industry attitudes. Costs of distribution of an added product may prove excessive. Costs of
liability need to be considered, in relation to control of timing
of degradation, and of disposal of degradation products,
expecially if enthusiastic claims of interest in large volume
markets (such as agricultural mulch) will be realized. This
newer technology may prove to be an effective antidote to selected
portions of the small (ca 6%) share of litter represented by
plastics. But will it be self-defeating if it creates in the mind
(?) of the litterbug license to litter more, since he will think it
will all self-destruct, though, in fact, not all materials, even
plastics, in litter are convertible to such degradable types?
Even worse, does not this characteristic frustrate recovery of
the resources represented by such plastics, either as material
or as heat? Clearly, carefully controlled evaluation must
be made to determine its real environmental cost/benefit.

GENERAL

The President's Council on Environmental Quality has made
it standard procedure to characterize any proposed Federal
action with significant effect on the environment. These actions
and their associated environmental statements cover a wide
latitude.

In the case of interactions of plastics, their components
and their products of combustion, with all aspects of the environment, including the human system, one has already been put in
this Federal format. This is the Internal Revenue Service's
Bureau of Alcohol Tax and Firearms statement on the environment impact of unrestricted use of poly(vinyl chloride) bottles
for alcoholic beverages. EPA has found this statement unsatisfactory and requested further information. Many other interactions are under study, but not yet expressed in this Federal
format.

We will review some of the more important interactions in
such diverse situations as food-packaging, medical devices,
materials of construction under fire conditions, and in solid
waste disposal. It will be evident that though there are frequent

allegations to the contrary, plastics provide a high benefit/risk
ratio for such environmental contacts and a capacity to contribute further gains.

FOOD PACKAGING

Some argue that the birthrate of new polymers for this
and other large-scale uses of plastics is dropping rapidly.
Others point to the development of newer polymers such as the
nitrile-based "barrier" types which, for the first time, offer
levels of impermeability comparable to those of traditional
glass and metal. Also, they offer far less weight per container,
and new options for solid waste management.

Limiting ourselves, for the moment, to the interaction of
these new polymers with the human system, and specifically to
proving safety of foods so packaged, we can point to changing
factors in such proof. There are pressures, Congressional and
otherwise, to include in the elements of safety, freedom from
mutagenicity and teratogenicity. These factors are involved in
Senator Nelson's and other bills to amend the Delaney carcino-
genicity clause of the 1958 Food Additives Amendment. An
industry view, with some support from FDA, is that "no-effect"
levels for each of these factors can be determined by current
methodology. This may be supported by the ultimate results of
the exhaustive studies of saccharin.

But other direct and indirect environmental factors are
involved. If the regulations on Public Information Policy are
adopted as proposed, FDA will insist on types and quantities of
information to prove the safety of incidental food additives from
food packaging materials which will satisfy its severest critics,
who will have access to such information. This can be trans-
lated to higher costs for additional analytical and toxicological
information, to support Food Additive petitions.

The long-rumored proposed regulation on colorants for
food-packaging has appeared. This will have the greatest effect
on those colorants not included in the regulations, those currently

used on the basis of their demonstrated lack of migration as determined by analytical methods of specified sensitivity. This will focus attention on the benefits and risks of traces of certain heavy metals, especially selenium and cadmium. The present proposal combined with certain local air-pollution codes effectively eliminates channel black and provides no alternative for this important and safe colorant for plastics.

MEDICAL DEVICES

The increasing use of plastics in bio-medical applications inevitably means greater concern for the interactions between plastics and the human body. So far the outstanding example is the agglomeration of platelets in human blood stored in plasticized PVC bags for the majority of the permitted storage period, 21 days at 40° F. Much of this story has been told by Dr. Barry Commoner in his book, The Closing Circle -- Nature, Man and Technology, but there are several facets which need to be put into perspective.

1. Perhaps the earliest investigator of this phenomenon, Dr. Robert Rubin at Johns Hopkins, is still engaged in defining the true mechanism by which phthalate plasticizers may affect blood.

2. One phase of Dr. Rubin's study is to determine if there are other sources of phthalates which may enter the human system by inhalation, injection or ingestion, and if so, what, if any, is the subsequent reaction.

3. Premature and inaccurate publicity of this work created a chain-like reaction of other reports of ubiquitous phthalates originating in nature and in man-made materials, automobile upholstery and thence in man. Petroleum, tobacco, milk, and other sources have been mentioned. Contamination of extracts prepared in laboratory equipment containing phthalate plasticizers has been noted. These proved to be a factor in early reports of the allegedly high phthalate content of blood of drug overdose victims. This is now recognized to have been exaggerated.

4. Dr. Commoner himself hinted ominously that phthalic
        anhydride is a constituent part of the now notorious
thalidomide. He did not refer to recent studies of specific
phthalimide pesticides proving these to be not teratogenic. These
latter do contain the phthalic anhydride moiety but not the glutar-
imide moiety of thalidomide.

5. Early this year Dr. Autian, of the University of
        Tennessee at Memphis, reported that eight phthalate
plasticizers, especially the lowest molecular weight types, may
be teratogens, in animal tests. Further work would be necessary
to determine such effects in humans. We note that half the
abnormalities cited for the 22 phthalates in rats occurred at
frequencies equal to or less than those for the four non-phthalate
treated controls.

6. Such reports, and others showing possible teratogenic
        effects of phthalates in fish have prompted further on-
going reviews of the entire toxicological pattern of phthalates,
with reference to common contaminants and residues from their
production. FDA plans a symposium on this subject next week
at the National Institute of Environmental Health Sciences at
Research Triangle Park.

7. Simultaneously, improved plasticizers, not all phtha-
        lates, with markedly reduced volatility, extractability,
and toxicity, due in part to greatly improved purity, are being
or have been made commercially available. This illustrates an
effective response of the plastics and component-supply industries
to correct a possibly undesirable environmental impact.

MATERIALS OF CONSTRUCTION
UNDER FIRE CONDITIONS

This subject is a very broad one and deserves separate
treatment. Suffice it to say that we appear to be on the threshold
of a much more sophisticated approach to the whole problem of
the interaction of burning materials of construction (not only
plastics) and the environment. Evidence for this is in the more

than 40 bills in Congress which would provide new authorities and monies for many aspects of fire prevention and control. The synthesis of a total national strategy is the most formidable responsibility of the National Commission on Fire Prevention and Control. Within one year, this small group must report its recommendations for this strategy, possibly requiring legislative and ultimately regulatory activity.

The plastics industry, through the Society of the Plastics Industry, is contributing major perspective and guidance to meet this goal for plastics. The following areas are being considered:

1. The actual growth of plastics as materials of construc-
        tion are not limited to structural products, i.e., pipe,
conduit, cable covering, floor tile, but also include furniture,
wall coverings, and probably also synthetic fiber in furnishings.

2. Definition of the several varieties of plastics, with
        descriptions of their properties, availabilities in
specific products, for the guidance of architects, designers and
others concerned with the selection of materials for the above
products and for residential and office construction.

3. Identification of performance of plastics under the full
        range of fire conditions, ranging from pyrolysis to high
draft. This includes products of combustion, their toxicities,
singly and in combination, with some emphasis on resistance to
ignition and performance during early stages of fire when escape
is still possible, as opposed to ultimate combustion conditions.

4. Conversion of this information to Model Building Codes,
        Insurance Codes, Recommended Practices for Ware-
housing, and to Standard Practices for Firefighters.

5. Identification of unanswered questions which will justify
        adequate research programs at appropriate institutions.
These would include, for example, improved early-warning instru-
mentation, combined flame- and smoke-retardants, and test
protocols more reflective of design, installation and performance
under real fire conditions than laboratory tests of test specimens.

With the plastics industry taking an active role in this substantial program, the recurrent fire-related difficulties with PVC, poly-(urethanes) and other materials should be reduced.

PLASTICS IN SOLID WASTE DISPOSAL

Plastics have received their worst press on pollution of the environment during their disposal. It apparently matters very little to the zealots that plastics amount to only 2 to 3% of all collected municipal waste, and 6% or less of land-borne litter, which itself is 6% or less of all collected municipal waste. This is hardly descriptive of a "plastic throw-away age", a phrase frequently used by anti-establishment people. Some point to experience in Japan, where very rapid post-war growth, an extreme shortage of paper, and great emphasis on plastics in all possible applications have helped increase their level of plastics in waste to 7 or 8% in Tokyo, for example. Others believe that the rapid growth of plastics in the USA will shortly find us repeating the trends in Japan. My view is that this is not likely.

The concerns for environmental impact of the disposal of plastics arises from their alleged increases of air and water pollution and deleterious effects on existing processes for disposal. Also, there is much erroneous talk about the resistance of plastics to recycling, frequently considered an alternate to disposal.

We can evaluate these concerns by analyzing the adaptability of plastics to existing and forecast methods of disposal, and of recycling. Nine to ten percent of collected municipal waste is incinerated, with various degrees of effectiveness. The remainder is simply dumped, with an increasing replacement of dumps by landfills, with a wide range of levels of sanitary quality. Those landfills operated by the larger cities (100,000 or larger in population, and accounting for 50% of the total USA load) are increasingly largely sanitary. Thus, in time, spurred by EPA's "Operation 5000" program, sanitary landfills will hopefully replace all dumps.

PLASTICS IN LANDFILLS

In sanitary landfills, plastics have proven highly compatible. Their very resistance to bio-degradability insures no production of acidic carbon dioxide from upper aerobic layers, and flammable methane, noxious organic sulfides and corrosive hydrogen sulfide from lower anaerobic layers, or of leachate which might contaminate ground waters. As a result, site settling is minimized. Los Angeles County, which makes heavy use of well operated sanitary landfill, reports no difficulty whatever with plastics. Other locations have evaluated or practiced grinding of waste prior to landfill, with no difficulties. This avoids the rare complaint of loose plastic film fouling truck or bulldozer radiators.

PLASTICS IN INCINERATION

Greatest concern is expressed about incineration. Many allegations have been made but the facts are these.

1. A survey of USA incinerator operators found <u>one</u> expression of difficulty and this was hearsay evidence of excess smoke. Most operators hoard plastics to improve speed or quality of "burn-out" of other combustibles, thus reducing smoke, especially during wet seasons. When incineration difficulties occur, they appear unrelated to any one type of material. The estimated national average of incinerator feed composition shows paper in all its forms to be the single largest component, 20 times the proportion of plastics.

2. Another analysis, by De Bell & Richardson, Inc., for the Manufacturing Chemists Association, Plastics Committee, found no evidence of the alleged corrosion of metal parts, grate clogging, or excessive stack emissions due solely to plastics. Although 70% of all USA incinerators are inadequate for "burn-out" (efficiency of combustion) and for air pollution control, due to poor design or operation, the difficulties encountered are not due to plastics, but related to all materials present.

3.   The one plastic material which EPA's Office of Solid
     Waste Management has related to incinerator corrosion
by "circumstantial evidence" is poly(vinyl chloride). It is true
that one pound of PVC resin does produce 0. 56 pound of HCl, but
this fact needs to be put into perspective, in terms of the small
amount of PVC in incinerated waste, the large amounts of other
sources of HCl, and of other acid-formers, and established
technology for removal of HCl from stack emissions, or their
dilution so that ground level concentrations are insignificant.
None of this perspective has been reflected in the spate of
(defeated) legislation as in the Michigan and New York State
Assemblies to outlaw PVC containers.

4.   Chlorine appears in paper, food wastes, yard waste,
     textiles, rubber and leather, as well as in plastics.
Paper accounts for 30-50% of the HCl produced, as do all rubbers,
leather and plastics. Poly(vinyl chloride) alone accounts for no
more than half this latter fraction of HCl. The fastest growing
source of HCl is textiles, due to the use of chlorinated flame-
retardants increasingly required to meet the demands of Federal
regulations. It has been estimated that all polymeric halogen
(principally chlorine) added for this purpose to plastics and
textiles was only 2% of all halogenated polymers, including PVC,
in 1969, but will quadruple to 8% in 1974. Furthermore, most
longer range projections show paper increasing past 50 to 55%,
plastics increasing slightly to 3 to 4%, with glass and metals
showing the major decrease. We should keep in mind that collected and incinerated municipal waste is a small part of the total.

A careful, more detailed analysis of chlorine and other
important acid formers in fractionated collected municipal waste
has just been started by the Society of the Plastics Industry, in
coordination with the National Center for Resource Recovery.
The methodology used is intended to permit repeated monitoring
of these factors.

5.   Studies were carried out at New York University's
     School of Chemical Engineering for the Society of the
Plastics Industry on a full-scale 1966 vintage incinerator at
Babylon, Long Island. They showed the harmless effects of

increasing each of four plastics components of municipal waste
by increments of 2 and 4% (absolute). The average percent distribution of types of plastics in packaging waste, as this is in
collected municipal waste, is known. The bulk of plastics (76.6%)
is in other than packaging uses, thus with longer sojourns
between use and disposal.

New York City's EPA has chosen to criticize this work, but
but with more emotion than fact.

Even poly(styrene), which burns in the open with much
black smoke, reduced smoke from the Babylon incinerator, due
to the slightly higher temperatures obtained from its higher BTU
content, an advantage exploited by many incinerator operators
during wet weather. Poly(olefins) behaved similarly. Poly-
(urethanes) and PVC showed insignificant increases in smoke.

Sampling a portion of the stack emissions for HCl showed
how this was increased over the normal background level. The
increase was disproportionately lower than the added amount of
PVC. This reflects the substantial amount of HCl from other
materials (as indicated above) and the combination of nearly half
the HCl with the incinerator residue. The laboratory scrubber
removing HCl from the sample stream was 86% effective,
approaching the performance of large-scale equipment.

No detailed study of metal corrosion was made since this
is the objective of a continuing series of analyses, especially
by Battelle, here and abroad. A definitive understanding of such
corrosion is not yet at hand. Maintenance of an oxidizing atmosphere and temperatures above the dew point and below 1000° C
seems necessary. Regardless, corrosion can be serious in the
absence of HCl and negligible in its presence.

One problem not yet quantified is the possible pollution of
air by particles of heavy metals by incinerating plastics containing
pigments and stabilizers based on cadmium, nickel, antimony,
lead, etc. As calculated, this is not an insignificant source, but
actual measurements need to be made. The efficiency of

stack-emission cleaners is adequate to remove these particulates, but the extent of their use needs to be increased.

PLASTICS IN COMPOSTING

So much for the compatibility of plastics with sanitary land-fill and incineration. Plastics are likewise compatible with composting, especially when reduced in particle size to form a useful inert fraction. Whether or not composting will become economically successful in this country remains to be seen. Twenty compost plants in the USA have been abandoned as uneconomical.

The All-American Environmental Control Corporation plant, recently started in Delaware's New Castle County, will provide a large-scale test of the feasibility of reclaiming domestic solid waste, roughly half as humus.

FUTURE PROSPECTS

What of the future? What will be the developing pattern of disposal of plastics, as part of the total mix of materials we manufacture, use, and discard? How will this pattern affect our environment?

We have indicated that open dumps are already yielding to landfills, stimulated by EPA's "Operation 5000" program. The Environmental Protection Agency indicates that short-term plans emphasize technical assistance for upgrading existing solid waste disposal systems, less so resource recovery. Economic and market analyses must be made to ascertain that markets can be developed for recovered materials. Public education needs to be greatly expanded. A few highly selective areas of new technology will be encouraged. But we believe the larger-scale use of "front-end separation", as urged by the National Center for Resource Recovery, will change the character of landfills from "catch-alls" to a much more highly selective use and recovery of materials. Franklin, Ohio, is perhaps the primary municipal example of such a wet-process separatory system, based on

Black-Clawson's Hydro Pulper equipment. This now reclaims
only cellulosic fiber but has "add-on" capability for other fractions when demand arises.

General Electric has proposed a similar but larger scale
program for New York State and for Connecticut. Low value
organic fractions, with some inevitable inerts, will be used for
deliberate reclamation of land on a wide scale. Such separatory
techniques will enrich high valued organic fractions for use as
fuel for power-generation purposes as is being evaluated at St.
Louis' Union Electric Company.

PYROLYSIS

With some degree of prior removal of non-combustibles,
pyrolysis (as destructive distillation, or perhaps more like
"starved-air" incineration) will probably gain in acceptance.
Initially, this will probably serve only as an alternative to incineration with lower costs for minimizing air pollution. In time,
the value of recovered chemical compounds may become important.

PLASTICS AS FUEL IN HEAT RECOVERY

Until very recently, fuel values of selected organic solid
waste fraction have been little recognized in this country. This
contrasts sharply with established practice in Western Europe
where 99 incinerators in 12 countries produce useful electricity,
or heat, for industrial purposes, residential heating or for drying
of sewage sludge. One of the newest, and certainly the largest
USA incinerator, is Chicago's 1600 tons/day unit. This exploits
the heat developed. It is not known if the planned 6000 tons/day
Kearny, New Jersey, unit will do the same.

This use of plastics may have the greatest environmental
benefit for lowest cost; in effect, it closes the ring by extract-
ing fuel value from plastics which originate as the products of
recovered volatile wastes from the petrochemical industry.

RECYCLING OF PLASTICS

The growth of recycling as a major factor will be slow.
Actually, many solid waste disposal officials consider recycling
efforts as <u>needless interference</u> with efficient disposal of refuse.
Some opportunities to sell off-grade plastics for low-cost hand-
sorting and reclamation overseas have been frustrated by AID
restrictions. Any viable national policy can not for long interrupt
international movement of plastics, paper or other material for
reclamation.

Recycling will become significant to the extent that, in this
country, private initiative pulls "re-used" or recycled materials
into useful and accepted products. This economic "pull" will
probably be "pushed" by various Federal economic incentives
such as Federal procurement policies and revision of tax structures such as the depletion allowance, which have had the effect
of over-stimulating use of virgin materials, and suppressing
recycling. The plastics industry has long shown the necessity of
recycling <u>in-plant</u> wastes to maintain profitability. What is
now necessary is to expand these techniques to rework consumer-
used wastes after adequate separation, by means now being
developed by the Bureau of Mines and elsewhere.

This turn-around will reflect some concern for dwindling
natural resources to be considered by the National Commission
on Materials Policy, but a greater impetus arises from direct
environmental effect. One general conclusion evident in "total
environmental impact" studies or "total energy demand" is that
large-scale systematic recycling can reduce the impact (or energy
demand) of any one material 50 or 60%. Largely for economic
reasons, this will clearly not occur overnight. It is anyone's
guess if large-scale entrepreneurial recycling will be in effect
in 30 years.

There is clearly a very wide range of attitudes toward recycling. The environmental zealots seem to endow recycling
with some of the virtues of perpetual motion, a panacea for <u>all</u>
problems of disposal and resource recovery. On the other hand,
the realist considers that if perfect control of each material in

all its products were permitted from start to finish of its life-
cycle, with an unheard-of high efficiency of 90%, only <u>six</u> recycl-
ings would dissipate, in loss or waste, 50% of the initial material.
(This assumes <u>no</u> degradation of the material!)

One charitable interpretation of most privately sponsored
community recycling activities is that they create awareness,
while large and more effective entrepreneurial facilities are
being developed. Clearly, recycling, stimulated as much as
possible by economic incentives, can make a significant contribution to an improved environment, but it is hardly a complete
answer. Recycling will require non-traditional economics and
investments.

It is frequently said that plastics cannot be recycled, as
opposed to glass (5%), paper (20%), metals (iron and steel, 55%;
copper, 50%; lead, 65%; aluminum, 25%) and textiles (30%).
(The high figures for metals reflect dwindling local natural
resources.) This charge is untrue. Thermoplastics, in particular, are highly amenable to reprocessing. The critical problem
is an ironic one. Plastics are so small in the total waste stream,
and so low in cost per pound, there has, so far, been no incentive
to separate out this highly dilute family of materials for recycling
on an economic basis. As has been pointed out elsewhere, as
bulk uses of plastics increase in number and in individual volume,
the problem and cost of the collection of once used plastics will
be eased, and the reconstitution of the recovered materials will
become a more viable and attractive proposition. These considerations should be put in sound perspective by an analysis and interpretation of plastics recycling being made for the SPI. The
report should be ready later this year.

CONCLUSION

Recently, a guest editorial in "Science" emphasized that
any crisis such as the environmental crisis leads to a frantic hunt
for new solutions. But saturation is soon reached and interest
melts because the limited attention span of the public is diverted
elsewhere. Then, away from the forefront of the media and

public attention the battle is fought in the face of waning interest
and disenchantment with quick solutions. The effort requires
hard, patient work aimed at gradual but real improvements. At
that stage, analysis and scientific inquiry make their most valuable contribution to solving social crises. That stage, I believe,
is only now being reached for Plastics and the Environment.

THE PLASTICS ISSUE

George L. Huffman and Daniel J. Keller

U. S. Environmental Protection Agency
National Environmental Research Center
Cincinnati, Ohio

INTRODUCTION

The Solid Waste Disposal Act of 1965 provided technical and financial assistance to State and local governments and to the private sector to do a better job of solid waste management.

The Resource Recovery Act of 1970 significantly amended the original Act, and the Environmental Protection Agency (EPA) was later given the responsibility of carrying out the dictates of this new legislation. In the new Act, Congress recognized that our affluent society was, in fact, quickly depleting its natural resources by allowing discarded matter, a potential resource of the future, to either go up in smoke or to be permanently entombed in our Nation's thousands of landfills. Clear, too, was the fact that large portions of valuable land in our urban areas were being used up in the process.

Clearly, there is a problem, and this Nation must direct its efforts toward a solution. Solid waste management, however, has many facets and our challenge today is to assess which aspects of the problem are most readily solved and, more importantly, which efforts will have greatest over-all impact in preserving the quality of our environment.

155

This paper is directed toward the disposition of one of the issues confronting our environment, The Plastics Issue. Its purpose is to analyze the environmental effects of plastics as they relate to solid waste management and to environmental quality.

Plastics are but one of the concerns of solid waste management. Others are paper, metals, glass, food wastes, the abandoned automobile, demolition materials, agricultural wastes, and industrial and mining wastes. Plastics have become a major concern of ecologically-minded persons: the obviousness of the plastic litter than remains on the countryside long after paper litter has decomposed, or the pitted corrosion of an incinerator after poly(vinyl chloride) (PVC) has been disposed of, or the possible health effect of hydrochloric acid discharged to the atmosphere as PVC is incinerated.

Background

Before discussing the effects of plastics on air pollution, incinerator corrosion, land disposal, litter, and recycling, the quantity and types of waste plastics involved and how these waste materials are being managed today should be outlined.

About 19 billion pounds of plastics were produced in the United States in 1969. The bulk, 75%, are the thermoplastics that are discarded when their useful service life is past. These 14 billion pounds of the major plastics (e. g., high and low density poly(ethylene), poly(propylene), poly(styrene) and the environmentally-famous PVC) will eventually enter the solid waste stream.

The total amount of plastics found in the waste stream in 1970 has been estimated at 6.5 billion pounds, or about 2% of the municipal solid waste load. About 70% of this waste plastic is poly(ethylene) and poly(propylene), 17% is poly(styrene), and the remaining 13% is PVC. Therefore, the PVC content of municipal refuse is about 0.2 to 0.3%. About 60% of these waste plastics (4 billion pounds) comes from packaging materials, about 15% from manufacturing/production operations, and the remainder from a variety of sources (housewares, 6%; toys, 5%; and wire and cable, transportation, appliances, furniture, etc.). [1]

Waste plastics are generally disposed of in much the same fashion as municipal solid waste and are normally mixed in with the municipal refuse load. That is, most is dumped openly, about 9 to 12% is incinerated, and about 10% is disposed of in sanitary landfill operations. Today, essentially no plastics are recycled from the waste stream. However, 15 to 20% of production is recycled within the plastic-fabricators manufacturing process and from resin producer to reprocessor.

With this brief introduction, let me proceed to some of the issues surrounding the disposition of waste plastics -- first the problem of plastics-caused air pollution.

INCINERATOR AIR POLLUTION

Several air pollutants are formed when various types of plastics are incinerated. For example, HCN can form when acrylic-type plastics are thermally decomposed; HBr can form from the bromine compounds added to some plastics as flame retardants. [2] In most cases, however, the amounts and concentrations of these air pollutants that result from the incineration of plastics are quite small, though the amounts may well be increased in the future. Considerably higher tonnages of another air pollutant result from the incineration of waste PVC plastic, namely, HCl. The HCl generated from incinerating PVC has received much coverage in the literature, and although it is indeed a significant incinerator air pollution problem, HCl is not one of today's major national air pollution problems. In terms of tonnages, HCl does not compare with such pollutants as SO_2, CO, NO_x, and particulates. In terms of emissions, considerably more HCl is emitted to the atmosphere from the Nation's coal-burning power plants than that which emanates from municipal incinerators. HCl does, nonetheless, represent an air pollution problem in the immediate vicinity of municipal refuse incinerators.

HCl emissions from all sources (including incineration) can be readily controlled by commercially-available flue gas scrubbing techniques. Scrubbing with dilute solutions of alkaline materials, such as NaOH, NH_4OH, $Ca(OH)_2$, and $Mg(OH)_2$, can remove bulk quantities of HCl, SO_2, and SO_3 from incinerator off-gases. Scrubbing with water alone can remove most of the HCl, SO_3, and particulate matter, with an alkaline scrubbing liquor needed for SO_2. The use of "water-only" scrubbers for incinerator effluents is fairly common; the use of alkaline liquors, quite uncommon.

Technology does exist, then, for the control of HCl emissions: were this technology applied extensively, PVC incineration would represent an extremely small fraction of the Nation's total air pollution problem. It should also be recognized that application of this technology will be costly and in many instances, communities may not have the necessary funds to upgrade existing incinerators. Further, many incinerators are so antiquated that they may not be worth upgrading.

Applying scrubbing technology to control incinerator HCl emissions is not trouble-free, however, since the problem of incinerator corrosion remains.

Incinerator Corrosion

Most present-day incinerators suffer from corrosion of metal surfaces. Although some of this corrosion can be attributed to the low concentration HCl resulting from the combustion of the PVC content of refuse, other corrosion-causing acid gases are present as well and these, too, contribute to the over-all corrosion. Even if plastics were completely removed from the incinerator feed, SO_2 and SO_3 in the flue gas from sulfur-bearing solid waste material and HCl from feed materials such as lawn clippings and grass would still cause corrosion.[3] The presence of HCl in the flue gas is nonetheless of particular concern because this material accelerates pitting of metallic construction materials. It also limits the use of austenitic stainless steels because of the possibility of stress-corrosion cracking.[4]

Most of the metal corrosion difficulties appear when the temperature of the flue gas is lowered below that of its acid gas dew point. This occurs when incinerator flue gas effluents are subjected to any scrubbing process whether the scrubber is installed simply for the control of particulate emissions or for the control of all emissions.

The appropriate choice of materials for the internal parts of the scrubber can minimize corrosion difficulties. A study conducted for EPA by the Battelle Memorial Institute of Columbus, Ohio, investigated the corrosion problems experienced at several incinerator scrubber installations. [4]

In their survey which paralleled actual field testing, Battelle found that in today's low-pH, water-only scrubber circuits: (1) carbon steels, cast iron, brass, and bronze are rapidly corroded; (2) performance of stainless steels has been quite variable, ranging from little corrosion to severe wastage and pitting; and (3) the more corrosion-resistant materials are the alloys (Hastelloy C, titanium, and Inconel 625)* which are quite expensive.

During their actual field tests at the scrubber sites, Battelle found that these alloys do, in fact, display superior resistivity to corrosive attack, and that they are resistant to stress-corrosion cracking. [5] The method used in the field was one of inserting several corrosion probes within the corrosive environment of the scrubber.

Battelle's study also pointed to the use of non metallics such as acid brick or Fiberglas-reinforced* plastic for the internals of incinerator scrubbers. These materials have reportedly shown good resistance to corrosion at the venturi scrubber installation at New York City's East 73rd Street incinerator. [6]

---

*Mention of a commercial product does not imply endorsement of the U. S. Government.

The induced-draft fans that are downstream of these scrubbers also experience corrosion and stress-corrosion cracking difficulties, primarily as a result of sulfate and chloride deposits on the fan blades. [7] Judicious choice of fan material is necessary here as well, unless more drastic steps are taken: locating the fan upstream of the scrubber or providing flue gas reheat capabilities.

LANDFILLING OF PLASTICS

In addition to incinerating plastics along with the rest of solid waste load, another method of plastics disposal is the sanitary landfill technique. [8] One often hears (1) that plastics should not be disposed of in a landfill because synthetic plastics, with very few exceptions, do not decompose significantly within the time frame of refuse decomposition and thus delay reuse of the landfill site; (2) that plastics at the landfill site are aesthetically unacceptable because the light film materials are blown across the disposal site; or, (3) that the resiliency of many plastics prevents their being well compacted, and thus they create voids in the fill that occupy valuable landfill space.

Permit me to discuss each of these claims. Because synthetic plastics are essentially non-degradable in the landfill environment, they will not add to the production of leachate or decomposition gases as do most other components of refuse. Thus, from the standpoint of water pollution and control of decomposition gas, the relative non-degradability of plastics is a <u>definite asset</u>.

Operators of sanitary landfills attach no major significance to routine handling of small quantities of plastic film included with municipal refuse. Large quantities of waste plastic film could, however, pose special handling conditions. In a properly operated sanitary landfill, movable litter fences placed downwind of the operating face can control blowing paper and film plastic. Litter beyond the working face can be picked up by hand near the end of each operating day. Film plastic <u>can</u> be a nuisance to the bulldozer operator as it -- and paper -- sometimes collect on the radiator grill. Reverse fans or periodic removal by hand can help control this.

Incorporating plastics into a sanitary landfill should scarcely have an adverse affect on the ultimate use of the landfill. Landfilled refuse decomposes very slowly, and so-called "stable" fill is, at best, only chemically and biologically stable; it certainly is not stable in the common sense with regard to its structural properties. Any concentrated load imposed on even a "stable" refuse fill will likely be subject to long-term and possibly differential settlement.

In the authors' opinion, attempting to relate landfill quality or ultimate use of a disposal site to just one of the waste ingredients is not proper, particularly when the ingredient constitutes only a small portion of the fill material. It is proper, however, to relate poor quality and inability to use the site to improper design and operation of the fill. This brings us to the basic issue. If a sanitary landfill is designed and operated with today's technology, plastics disposed of in the sanitary landfill should pose no special problems to the operation or to the ultimate use of the site.

In summary, operators of sanitary landfills experience no significant problems when handling municipal refuse containing plastics. Operators can compact rigid and flexible plastics using proven, conventional techniques of spreading in thin layers. Resiliency is no problem. Once crushed, the mass of overlying refuse or soil can hold the plastic object in the flattened condition the same way an overlying mass can keep brush and tree clippings compacted. The key to the issue, then, is not the ingredient but how the ingredient is placed in the fill. If properly placed, plastic should occupy no more "valuable" landfill space than other non-degrading materials.

PLASTICS AND THE LITTER PROBLEM

Everyone is exposed to litter along our highways, in our parks, and along streams and beaches. Littering is a highly visible and serious problem of solid waste management, and plastics are an obvious contributor. If predictions of increased plastics use come true -- and people do not change their habits -- plastics will become an even greater contributor to the litter problem.

Notwithstanding the anti-litter laws of most states and most communities, the numerous advertising campaigns conducted by industry, government, and private groups to encourage people not to litter, and the provision of more litter containers than ever before, the national disgrace of littering continues.

One approach to help solve the litter problem is modifying the package so that it will, when littered, eventually "self-destruct". Thus, a water-soluble glass container and a degradable plastic container are being developed. A number of questions about the performance of these materials remain: their safety, their durability, their susceptibility to disintegration once discarded, and the nature of their disintegration products. Sociological/psychological questions arise, too. For instance, since many citizens are already callous about littering, would self-destroying containers provide the litterer even more of a license to be careless?

Meaningful statistics on litter quantities and composition are quite sketchy, and the actual cost of litter control across the Nation is very difficult to estimate. "Chemical Week", December 8, 1971, reported that in one study roadside litter contained only 6% plastic and that 82% was paper, metal, and glass (60, 16, and 6%, respectively).

More litter laws, enforcement, improved methods of litter pickup, and design and use of "people-proof" self-destruct packaging might someday become effective, but it seems that the real long-term solution for control of all litter is proper citizen attitude.

PLASTICS RECYCLING

An ideal solution to our Nation's mounting solid waste
problem lies in increased recycling. The secondary materials
industry continues to do a commendable job of recovering millions
of dollars of material from industrial and commercial sources
that might otherwise become part of the refuse stream. Technology has been developed to recover ferrous metals from the
municipal refuse stream, but separation of other constituents is
more difficult.

We have not demonstrated the technical capability to recover
economically plastics from mixed refuse. In recent years,
numerous approaches have been proposed, but to date, none have
been fully developed. Separation of the various individual plastics
from a waste plastic mix, while by no means an easy task, seems
to be a simpler problem to solve than separation of the plastics
mix itself from the other components of municipal refuse. The
Bureau of Mines, in pilot-plant studies, have met with success
in separating the individual plastics by utilization of flotation
techniques. [9]

Solution of the basic problem of primary separation of the
plastic mix from the other components of municipal refuse is,
at present, the key to increased direct recycling and reuse of
plastics.

Although this primary separation problem is a chief technical deterrent to the direct, as-is, reuse of plastics, work is
underway that would negate the need for accomplishing this
separation in the first place. That is, it is possible to derive
benefit from applying recycling technologies to the total municipal
refuse stream. Examples of these recycling techniques and processes and of the R & D being conducted are:

1. To recover the heat given off during the incineration of
        the plastics-containing solid waste, either as electricity
or as steam for heating purposes. An example here is EPA's
demonstration grant with the City of St. Louis in which refuse
is combusted right along with coal in a coal-burning power plant.

Another example is EPA's research contract with the Combustion Power Company of Menlo Park, California, in which combustion off-gases from a 90 ton per day, pilot-scale pressurized combuster are expanded through a turbine directly to produce power. [10]

    2. To recover the products of a refuse pyrolysis operation either as a pipeline gas or as feed material to an adjacent refinery. Examples here are: EPA's research grant with West Virginia University in which refuse pyrolysis is being studied on a bench-scale; [11] and the Bureau of Mine's research on refuse conversion to pipeline gas. [12]

    Recycling processes such as these are being developed in the hope that they will lead to full-scale commercialization of techniques that feature not only waste reutilization but also attractive over-all economies for the country.

    Another major deterrent to the direct recycling and reuse of waste plastics is economics: virgin plastics are relatively inexpensive.

CONCLUSION

    In conclusion, then, it can be said that:

    1. Technology exists for controlling the HCl emissions that result from the incineration of waste PVC plastic; application of this technology will undoubtedly result in increased cost; if this technology were applied on a wide scale, the contribution of plastics incineration to the Nation's total air pollution problem would be quite small; if this technology is not applied, the plastics contribution to air pollution will surely increase because of the projected increases in the usage of plastics.

    2. Though HCl and SO_3 scrubbing technology is commercially available, its application results in troublesome corrosion problems; these corrosion difficulties can be substantially reduced by selection of appropriate materials of construction

for scrubber and fan internals, but increased costs for this equipment can be anticipated.

    3. If appropriate sanitary landfill techniques are utilized, waste plastics in the landfill pose no special problems concerning operation or ultimate site resue.

    4. The long-term solution to the litter problem, in general, and thereby to the plastics litter problem, too, is citizen attitude.

    5. The chief technical impediment to the direct reuse of waste plastics is the separation of the plastics mix from the conglomerate of other components present in municipal refuse; present economics are also unfavorable to direct plastics reuse. Research and development work is underway to demonstrate recycling processes that indirectly utilize waste plastics in municipal solid waste. This includes incinerator waste-heat recovery and pyrolysis.

REFERENCES

1. J. Milgrom, "Incentives for the Recycling and Reuse of Plastics", Proceedings of the Solid Waste Resources Conference on Design of Consumer Containers for Re-use or Disposal, May 12-13, 1971, EPA Publication No. SW-3p, pp. 69-93.

2. M. E. Fulmer, "The Role of Plastics in Solid Waste", Battelle-Columbus for the Society of the Plastics Industry, 1967, p. 8.

3. A. J. Warner et al., "Solid Waste Management of Plastics", DeBell & Richardson for the Manufacturing Chemists Assn., December 1970, p. 1.

4. P. D. Miller et al., "Corrosion Studies of Municipal Incinerators", Battelle-Columbus for the Environmental Protection Agency, Solid Waste Research Division, 1972, p. 85.

5.   P. D. Miller et al., ibid., pp. 84-110.

6.   P. D. Miller et al., ibid., p. 111.

7.   P. D. Miller et al., ibid., p. 100-110.

8.   D. Brunner and D. J. Keller, "Sanitary Landfill Design
      and Operation", EPA Publication No. SW-65ts, 1971.

9.   J. L. Holman et al., "Bureau of Mines Technical Progress
      Report: Processing the Plastics from Urban Refuse",
      TPR 50, February 1972.

10.   D. A. Furlong et al., "Combustion of Municipal Solid Wastes
      in a Fluidized Bed", paper at the American Institute of Chem-
      ical Engineers National Meeting in Cincinnati, May 16-19,
      1971.

11.   R. C. Bailie et al., "Gasification of Solid Waste Materials
      in Fluidized Beds", paper at American Institute of Chemical
      Engineers National Meeting in Cincinnati, May 16-19, 1971.

12.   H. F. Feldmann et al., "Pipeline Gas from Solid Wastes",
      paper at American Institute of Chemical Engineers National
      Meeting in Cincinnati, May 16-19, 1971.

SUGGESTED ADDITIONAL READING

N. L. Drobny et al., "Recovery and Utilization of Municipal Solid
Wastes", Battelle-Columbus for the U. S. Environmental Protec-
tion Agency, Solid Waste Management Office, EPA Publication
No. SW-10c, 1971.

M. E. Banks et al., "New Chemical Concepts for the Utilization
of Waste Plastics", TRW Systems for the U. S. Environmental
Protection Agency, Solid Waste Research Division, EPA Publication No. SW-16c, 1971.

K. Gutfreund, "Feasibility Study of the Disposal of Polyethylene Plastic Waste", II T Research Institute for the U. S. Environmental Protection Agency, Solid Waste Research Division, EPA Publication No. SW-14c, 1971.

H. H. Connolly, "Plastic Wastes in the Coming Decade", address to the Society of Plastics Engineers, Cherry Hill, N. J., October 1970.

J. E. Potts, "An Investigation of the Biodegradability of Packaging Plastics", Union Carbide Corp. for the U. S. Environmental Protection Agency, Solid Waste Research Division, April 1972.

THE FEDERAL APPROACH TO RECYCLING OF POLYMERS*

John P. Lehman

Office of Solid Waste Management Programs
U. S. Environmental Protection Agency
Washington, D. C., 20460

Previous speakers have discussed the <u>technological</u> aspects of polymer waste management. I would like to turn your attention now to <u>policy</u> and <u>economic</u> considerations related to the recovery of resources from polymers. Unless otherwise stated, the discussions reflect my own personal views, and are not the policy of the U. S. Environmental Protection Agency (EPA).

The complexity of modern civilization, and our heavy utilization of chemical substances normally absent from a natural environment, are slowly producing social, political, and administrative changes which either have had or are likely to have impacts on industry's way of doing business.

The three most important aspects of this phenomenon of a changing regulatory climate are those of public health, safety, and the environment. In the environmental area, the Federal impact of air and water pollution regulations dealing with the so-called economic poisons -- pesticides, fungicides, and rodenticides -- and those dealing with detergents, are already well-known and, not surprisingly, have sensitized industrial decision makers. Today even rumblings of new regulatory initiatives by the Federal

---

* This document has not been formally released by EPA and should not be construed to represent EPA policy.

Government are enough to excite considerable interest. Following the lead of the Ford Motor Company, industry has begun to
listen better, and where a decade ago regulatory initiatives in
their embryonic stages were not taken seriously, today it is
enough to mention a concept at a cocktail party and inquiries
begin to flow in the next day.

Given this general mood, industrial interest in possible
legislative or regulatory actions by the Government related to
solid waste management is high, and is one reason, I suppose,
why this talk was scheduled in the first place. I am thus alert
to a basic problem any spokesman is likely to have: I do not
want to start any rumors, but, at the same time, I would like to
give you a reasonably accurate assessment of the subject.

LEGISLATIVE MANDATE

Today EPA's Office of Solid Waste Management Programs
has no legislative authority to regulate any aspect of industrial
or municipal solid waste handling or disposal. The Office has
authority to issue guidelines which are binding upon other Federal
agencies; otherwise, adherence to the guidelines is a voluntary
matter unless made mandatory by a lower level of government.
EPA does have jurisdiction over the water and air pollution
aspects of waste handling activities -- such as effluents from
incinerators and the generation of groundwater pollutants from
landfills.

Similarly, the Federal Government has no authority to force
anyone to recycle any waste material nor the power to prohibit,
ban, or restrict the production, distribution, and use of materials
or products in order to make waste handling or recycling easier.

The Government does have purchasing power and can use it
to buy products made of recycled materials and to cause disposal
of Government-generated wastes in a manner felt to be acceptable
to EPA.

This general situation describes the status quo. However, we are engaged in a dynamic process of social change and the status quo is likely to change, with the change being generally in the direction of incentives to make the recovery of waste materials more attractive to industry.

POLICY PRINCIPLES

Before turning to discussions of particular EPA program thrusts in this area, I believe it would be helpful to outline some general policy principles relating to solid waste management which were presented recently at Senate hearings by Samuel Hale, Jr., EPA's Deputy Assistant Administrator for Solid Waste Management Programs.

The Environmental Protection Agency believes that solid waste policy actions, particularly financing and regulatory actions proposed for implementation at the Federal level or recommended by the Federal Government for implementation at the State or local level, should be carefully measured and judged against certain criteria or principles before being acted upon in either a positive or negative fashion. These principles are outlined below.

I.   The environmental, economic, and social benefits of any policy should be equal to or greater than the environmental, economic, and social costs of the policy, if implemented.

The objective of this principle is to ensure that:

• environmental gains will actually result from particular policy actions;
• national resources are properly allocated in pursuit of environmental and other national goals;
• adequate accounting and hence evaluation of benefits and costs occurs before policy options are selected for implementation.

II.   The effect of a proposed policy should be reasonably predictable.

Although this principle is closely related to the first, it is necessary to ensure that:

• policy actions are based on adequate knowledge and data;
• policy actions, if implemented, will actually result in desired
effects;
• potential secondary or side effects, some of which might be
undesirable, are identified and carefully evaluated in light of a
particular policy's total benefit-cost situation.

III.   The generator of solid wastes should pay the full cost of
       collecting and disposing of his wastes without damage to the
       environment.

The objectives of this principle are to:

• ensure equity in the distribution of the costs associated with
environmentally sound solid waste management;
• create positive incentives to avoid unnecessary utilization of
solid waste services.

IV.   Policies with financing aspects should impact on solid waste
       systems in a positive, beneficial manner.

Stated more specifically, the purpose of this principle is to ensure
that financing policies:

• create incentives to optimize solid waste management system
productivity;
• alter producer or consumer behavior along preferred lines or
not at all.

V.   Policies selected should be administratively simple to
       implement.

The purpose of this criterion is to enhance changes for achieving
policy goals by minimizing potential administrative obstacles.

       Two further principles that logically derive from those
stated above should also be considered. Other things being equal:

VI.   A policy alternative is to be preferred if the success of its
       operation depends on natural economic and social incentives
       rather than on administrative judgment and action.

VII.   Broad market approaches to behavior alteration are, as a
       rule, preferred to narrow product-by-product approaches.

       In my view, these general principles are valid, and deserve
serious consideration in the policy development process. The last
principle has particular applicability to resource recovery.

                    RESOURCE RECOVERY THRUSTS

       Under the Resource Recovery Act of 1970, EPA has under-
taken a series of investigations and studies of resource recovery.
EPA's investigations are still in progress, and no specific conclusions have been reached as yet. However, a few general
requirements for National action to bring about resource recovery
have been identified.

       The high and growing rate of materials utilization in the
United States -- equivalent to 28 tons per capita in 1971, up from
22 tons in 1965 -- emphasizes the need for resource recovery as
a way to reduce our dependence on virgin materials resources,
to minimize the adverse environmental effects of materials usage,
and to reduce the quantities of solid wastes that must be handled
for disposal.

       While recovery of waste materials is a desirable alternative
to materials-processing using virgin natural resources, recovery
is declining relative to total materials consumption for a number
of reasons, all of which translate into adverse economics of
resource recovery.

       Artificial economic advantages favor the use of virgin mate-
rials by industry. These include depletion allowances, favorable
capital gains treatment, apparently discriminatory freight rates,
and not internalizing the full cost for controlling adverse environmental effects, i.e., certain air, water, and solid waste pollutants
are not yet subject to control, and, therefore, control costs are
not yet reflected in production costs.

EPA studies indicate that major total environmental benefits are associated with recycling of major materials (steel, paper, glass), when compared with virgin materials extraction and processing. Less air and water pollution, mining and solid waste generation, and energy consumption are associated with products made from secondary materials than with products made from virgin resources in these materials categories and, presumably, in others as well.

Recovery of materials from mixed municipal solid wastes, which is conceptually the best alternative to disposal, cannot be instituted on a large scale in the absence of (1) a substantial reduction in processing costs or upgrading the quality of recovered material, which are simply unattainable given reasonable projections of technology, or (2) a major reordering of relative virgin and secondary material prices to make the secondary materials more economically attractive.

STATUS AND TRENDS OF PLASTICS RECYCLING

Plastics are becoming an ever more important material in our society as their growth rate continues at an impressive rate. In the last decade, plastics consumption increased at an average annual rate of 12%, and totalled 8.5 million tons, in 1969. Consumption by 1980 is expected to reach 30 million tons. [1]

Today, plastics account for only about 1% by weight of municipal solid waste and by 1980 will average about 2%. Very little plastic scrap is recycled other than that reused within the manufacturing plant in which it is generated. This, however, is a fairly significant quantity. Plastics fabricators, for example, consume internal scrap equal to about 750,000 tons in 1970. [2]

The plastics reprocessor is the recycling channel for all industrial plastics recycled outside of originating plants. About 500,000 tons of waste plastics were handled by reprocessors in 1970. Of the plastics recycled through reprocessors about 55% came from resin products, 30% from fabricators and 15% from converters. [2]

There are two types of plastics, thermoplastics and thermo-
setting plastics. The thermosetts (20% of plastics consumption)
cannot be softened and reshaped through heating and are thus not
recyclable. In addition, most of the plastics used as coatings and
adhesives are impossible to recycle. Thus about 75% of the
plastics consumed are potentially recyclable.

Table I shows the major markets for plastics. Packaging
and construction are by far the most significant, accounting for
20 and 25% respectively of consumption in 1969. Plastics from
packaging account for about 60% by weight of the plastics in the
solid waste stream (much of the other plastic consumed is "held
up" in permanent and semi-permanent end uses). Although some
of the waste generated in the various stages of plastics production

TABLE I

CONSUMPTION OF PLASTICS, 1967 TO 1969, TOTAL AND
SELECTED MAJOR END USE MARKETS, IN 1,000 TONS*

Consumption	1967	1968	1969
Totals	6,550	7,558	8,535
In selected markets			
Agriculture	75+	85	95
Appliances	198	238	234
Transportation	109	334	536
Construction	1,070	1,215	1,327
Electrical	396	452	567
Furniture	250+	273	328
Housewares	313	373	425
Packaging	1,121+	1,508	1,729
Toys	208	243	269

*Source: A. Darnay and W. E. Franklin, "Salvage Markets
for Materials in Solid Wastes", Washington, D. C., U. S.
Government Printing Office, 1972, p. 88-5.

is recycled, the portion that is not makes up about 15% of the plastic in the waste stream. Thus, packaging and industrial waste account for 75% of plastic waste. [2]

A comparison of plastics recycling versus recycling of other major materials in 1967 is given in Table II. Roughly 5% of plastics consumed are recycled outside manufacturing plants compared with an average of 25% for all major materials taken together.

As a general rule, scrap plastic has to be used in an end application having wider specification requirements than the product yielding the scrap. The primary markets for scrap plastic include such items as hose, weather stripping, toys, cheap housewares, pipe, and similar applications where (1) plastics properties and performance are not paramount, (2) relatively noncritical

TABLE II

RECYCLING OF MAJOR MATERIALS, 1967*

Material	Total consumption, million tons	Total recycled, million tons	Recycling as % of consumption
Paper	53.110	10.124	19.0
Iron & steel	105.900	33.100	31.2
Aluminum	4.009	.733	18.3
Copper	2.913	1.447	49.7
Lead	1.261	.625	49.6
Zinc	1.592	.201	12.6
Glass	12.820	.600	4.2
Textiles	5.672	.246	4.3
Rubber	3.943	1.032	26.2
Plastics	6.550	.350	5.3
TOTAL	197.770	48.458	24.5

*Source: A. Darnay and W. E. Franklin, "Salvage Markets for Materials in Solid Wastes", Washington, D. C., U. S. Government Printing Office, 1972, p. xvii.

processes are used such as compression molding or heavy extrusion, and (3) the cost of plastic resin is a high proportion of total
product cost.

BASIC PROBLEMS

Some of the basic problems in plastics recycling are listed
below.

Technology. There is a fundamental difference between the
nature of plastics recycling and that of metals, paper, glass, and
other materials. Metals production, for example, begins with an
impure ore which is progressively concentrated, smelted, refined,
and freed from impurities. Plastics production, on the other hand,
begins with high purity virgin polymer to which various additives,
colorants, and reinforcements are added. Thus, in the metals
industries, there is a background of technology designed to purify
and upgrade ores and concentrates. Such technology can also
be applied to the upgrading of scrap. In the plastics industry,
where the basic raw material is progressively "contaminated" in
production, little technology has been developed which can be
applied to purify waste plastics.

Compatibility. The principal difficulty in recycling plastics
is that different polymers (poly(ethylene), poly(vinyl chloride), etc.)
are not compatible with each other and must be separated, a very
difficult and costly task.

Economics. Continually decreasing cost of basic plastic
materials has made scrap plastic less competitive with its main
competitor, off-grade virgin resin. For example, since 1961,
the price of low density poly(ethylene) has decreased from 24 cents
to 13 cents per pound. Scrap plastic, limited by rising labor and
distribution costs, did not drop as rapidly, and the price of the
scrap is now only about 1 cent per pound under the offgrade resin
price, versus about 3 cents in 1961.

Logistics. This problem, common to recycling of all materials, is important to plastics recycling. The extremely low density
of plastics makes transportation very costly.

Separation. Separation of plastics from the solid waste stream is extremely difficult, making their recovery from municipal waste almost impossible unless the plastics can be diverted from the waste stream and kept separate.

RECYCLING STRATEGIES

Given the general principles governing solid waste policy actions, the general thrust of resource recovery actions, and the specific facts and problems regarding plastics recycling, the next logical step is to outline and discuss the options available to promote recycling of plastics. Before considering such strategies, it is essential to examine some implicit premises and basic findings which underlie the discussion.

First, it is assumed that the cost to society of failing to clean up the environment is greater than the cleanup costs. In other words, the cost in terms of environmental damage of indiscriminently discarded plastics is greater than their effective disposal. It is also assumed that the social cost of effective disposal of a potentially recyclable plastic product is higher than the cost of recycling.

A fundamental finding of research to date is that systematic interdependencies exist in our ecological and economic systems. These interdependencies make any piecemeal action an inefficient and arid approach. Discouraging the use of any one substance (e. g., virgin plastics) may merely transfer pollution from one source to another. This implies that strategies developed for promoting recycling of wastes must include all materials (plastics, glass, metal, paper, etc.).

Incentives vs. Regulations

In a non-planned "mixed" capitalistic enterprise system such as ours, there are basically only two ways to control or condition production and consumption: (1) regulation, and (2) economic incentives or selected taxation. The essence of our system is consumer sovereignty, and the belief that production and consumption

decisions can best be guided by a price mechanism. Of course,
for a resource allocation to be efficient (that is, produce the correct decisions), prices of products and services in the market-
place must reflect their true social cost. To use enforcement to
promote recycling of a number of selected plastic products is inconsistent with the philosophy of a free market economy. Controls
should be used preferably when the consequences of certain acts
are outrightly prohibitive (consumption of poison, for example), or
when administrative impracticality renders taxation unfeasible.
We believe that under a scheme of appropriate economic incentives, a balance can be achieved between the objectives of (1)
minimizing environmental damage, (2) minimizing the cost of
recycling and disposal, and (3) minimizing economic disruption.
Accordingly, the strategies discussed below are built primarily
on economic incentives and disincentives rather than regulations.

Strategies

In order to bring about resource recovery at a high rate, the
fundamental requirement is to create a situation wherein industrial
materials users will substitute secondary materials for virgin
materials. This situation can be brought about by three types of
activities: (1) actions to inhibit the use of virgin materials, (2)
actions to create a demand for secondary materials, and/or (3)
actions to create a supply of secondary materials of such quality
and at such price that they will appropriately satisfy the new demand.

Among the more important action alternatives to inhibit
virgin materials use are: virgin materials taxes and/or the removal of favorable tax treatment of virgin materials and energy
substances; regulation of virgin materials that are available from
Federal land; denial of markets to virgin materials through
Federal procurement policies; changes in transportation costs
through Federal regulation of rail and ocean freight rates; and
the institution of national materials standards that would limit
the use of virgin materials in major materials. At a minimum,
increased Federal procurement of materials and products that contain secondary materials and changes in freight rates are clearly
desirable actions with forseeable beneficial results for resource
recovery.

To create demand for secondary materials, the most efficient approach appears to be to provide a positive economic incentive to the potential user of the waste materials by investment tax credits, tax credits for use of secondary materials, subsidy payments or bounties, subsidy of plant and equipment for processing or using secondary materials, and combinations of these. The most efficient incentive will be one that is available to the waste user directly, inducing him to exert a demand for the secondary material in order to qualify for the incentive. If "demand pull" is thus instituted, the various suppliers of the secondary materials will respond to the technical requirements of the user and will have an assured market for the waste-based commodities. Incentives for demand creation are veiwed as the first priority requirement to bring about resource recovery at high rates, and if properly designed, obviate the need for supply creation actions.

What should be clear from the general trend of these discussions is that:

• the preferred approach is not to regulate recycling of polymers, but to provide and economic incentive approach, and that
• specific incentives applying only to polymers are not appropriate since plastics are such a small fraction of the total waste stream. Recycling of polymers will most likely come under the umbrella of a more general waste recycling incentives program.

Incentive concepts will take some time to analyze, present, discuss and translate into effective legislation, possibly as much as two years, although we are hoping for quicker action. After the institution of incentives, some time will pass before the effects of the incentives will be felt.

REFERENCES

1.   A. Darnay and W. E. Franklin, "Salvage Markets for Materials in Solid Wastes", Washington, D. C., U. S. Government Printing Office, 1972.

2.   J. Milgrom [Arthur D. Little, Inc.], "Incentives for Recycling and Reuse of Plastics", U. S. Environmental Protection Agency, for distribution by the National Technical Information Service, 1972. (in press)

WHO'S ON THE CLEAN-UP CREW ?

Ralph J. Black

Office of Solid Waste Management Programs
U. S. Environmental Protection Agency
Washington, D. C., 20460

Most of our citizens are quite familiar with the solid waste collection system that they use. Few, however, really know what happens to the wastes after they have been collected in their community. And it is the rare individual that actually has visited the disposal facilities used by his community to personally judge the character and environmental effect of these operations.

The importance of citizen knowledge and support was recognized by the U. S. Environmental Protection Agency in the brochure, "A Citizens Solid Waste Management Project: Mission 5000".[1] As a part of this effort, citizens are encouraged to:

"Go and see how solid wastes are disposed of in your community.
Determine if your community's methods meet acceptable standards.
If not, ask local officials what special problems are preventing your community from adopting modern disposal techniques. Offer local officials your support. They need it."

This practical advice is also useful for the engineer or scientist who, because of his technical background, can exert real influence on local political leaders.

Assuming that you are an environmentally concerned citizen who wants to help improve solid waste management in your community and to protect the environment, what about plastics?

The basic trend that characterizes plastics use is rapid expansion. While the use of poly(vinyl chloride) in packaging and other products has been growing, it is only a small percentage of total plastics use. This has resulted in some industry spokesmen taking the position that PVC and other plastics are not a "problem" as far as the environment is concerned.

Similar positions have been taken by the spokesmen for the producers of every other packaging material. Taken individually, there is supportable logic for each contention, but in total, one would then erroneously conclude that there are no environmental problems associated with our present solid waste disposal practices. With 94% of the Nation's disposal facilities reported to be unsatisfactory by the States in 1968, [2] this is obviously not the case.

A growing number of manufacturers have taken the first step of making disposability one of the factors in product design. While the consideration of technical feasibility is basic, it is passive in its effect on solid waste management practices. The consumer is still encouraged to define "disposable" as being cheap enough to throw away after only one use.

In our present economy most companies use sophisticated production, distribution and sales systems to market their goods. These companies assume that any residues their consumers have left after enjoying the goods will be handled satisfactorily by the consumer's local solid waste management system. The plain facts are that most consumers don't have satisfactory local service.

It is significant that the 1968 national survey showed that approximately 12% of the residential population receive no formalized collection service and that another 11% receive only partial service. In addition, 14% of the population is served by systems that require separation of wastes at the source, yet their disposal systems have been combined. Over 12,000 land disposal sites were reported utilized by collection agencies and firms, yet 94% of these operations were unacceptable to the States. In addition there are approximately 300 municipal-sized incinerators, 70% of which were without adequate air pollution control devices.

Therefore, manufacturing companies, particularly those in the packaging industry, could best serve their own interests by taking an active part in efforts to improve local solid waste management systems. This can most effectively be done in the communities where company facilities are located. Most companies have materials-handling experts and other specialists who can offer real technical assistance and effective political support for needed improvements in local systems in their company's home town. Because companies are major taxpayers in their home towns, such interest can hardly be ignored by local officials.

Now, with the evidence of our past errors piling up around us, a new concept of solid waste management is emerging. It assumes that a real, workable system for managing our solid wastes must be devised by making the necessary changes in both the social and economic spheres. This involves:

- Controlling the quantity and characteristics of wastes;
- Recycling those that can be reused;
- Collecting and processing efficiently those that must be removed;
- Disposing properly of those that have no further use.

Your assistance in supporting these principles of effective solid waste management in your home town will help protect the environment while better serving your company and your customers.

REFERENCES

1.   Mission 5000; A Citizens Solid Waste Management Project,
      U. S. Government Printing Office, Washington, D. C., 1972.

2.   R. J. Black, A. J. Muhich, A. J. Klee, H. L. Hickman, Jr.,
      and R. D. Vaughan, "The National Solid Wastes Survey;
      An Interim Report", U. S. Dept. Health, Education and Wel-
      fare, 1968. Reprinted in Proceedings of the Third Annual
      Meeting, Institute for Solid Wastes, Miami Beach, Oct.
      1968. Chicago, American Public Works Assn.

PANEL DISCUSSION

EDITOR'S NOTE

On Tuesday, August 28, 1972, following the presentation of the technical communications, a panel discussion was held at which questions were directed to the authors of papers presented during the symposium. The discussion was chaired by Dr. David M. Wiles of the National Research Council of Canada. Arrangements had been made to have the entire session recorded on tape, but unfortunately the microphones used by the persons asking questions did not register on the tape with sufficient clarity to permit accurate transcription. In some cases the answers were also garbled. However many of the issues raised during the discussion were not dealt with in the formal presentations of the symposium, and I believe that the material covered is of sufficient general interest that it should be included in a volume of this kind. Since it was impossible to transcribe the questions exactly, I have taken the liberty of paraphrasing them in the text which follows. The answers, which were transcribed directly, are included with only punctuation and grammatical changes where necessary. I have included all of the discussion where it was possible to obtain a good transcription from the tape.

J. G.

QUESTION. How will the inclusion of photosensitive groups in polymers affect the possibility of recycling?

GUILLET. Photodegradation occurs only with that portion of the product which is not collected, i.e., litter. If you collect it as garbage, it will normally not have been exposed and hence can be recycled by whatever process is used for other plastics. Since there is presently no operable method for recycling plastics from garbage it is difficult to say whether or not it would be affected by including photodegradable plastics. We think it unlikely that any high value use can be made of plastic scrap recovered from garbage and that probably the most useful recycling will involve the recovery of the thermal energy by combustion and conversion to electric or thermal power. In this case there would be no measureable effect on recycling possibilities.

SCOTT. I would just like to confirm in our own experience the general philosophy that Professor Guillet has put forward. We have done quite a considerable amount of work on reprocessing and we've found that the sort of chemical changes that are introduced during the reprocessing operation are much more detrimental to the reprocessed products than are any additives which are included to make the material photodegradable. I would imagine that the same is also true for Professor Guillet's materials because it only required very small amounts of trace metal ions, for example, to completely change the nature of the polymeric material. I think that it must be realized that the whole problem of reprocessing is associated first of all with being able to collect the material cheaply and secondly with being able to purify it to a state which allows it to go through a reprocessing cycle.

QUESTION. Is it possible to restabilize additive systems for reprocessing?

SCOTT. We have done this in fact in the system which we have developed, which is a combination of a metal ion and a complexing agent. It is necessary to restore the amount of the complexing agent so that if metal ions are adventitiously introduced they are recomplexed back to the state where they are again anti-oxidants rather than the pro-oxidants produced during the UV exposure

procedure. So the answer to your question is, as long as the
plastic scrap has not been subjected to the UV degradation process
(and of course this would not normally be returned into a waste
disposal system anyway), then there is no reason whereby the
judicious addition of a small amount of the complexing agent should
not allow you to reprocess to a material which is very similar to
the original material.

GUILLET. In our system, unless the material has been extensively
exposed to ultraviolet light, there would be no problem whatsoever.
A certain amount of degradation is obtained with any polymer as
you pass it through an extruder, which you have to allow for when
you are reprocessing or adding scrap to a material. In our experience, the change that you get from working the material in the
extruder is greater than any change that might result from exposure
to the lights of the factory in which it is being processed.

QUESTION. Please comment on your additive in the area of use
of agricultural mulch films. Are the projected figures for use of
photodegradable agricultural mulch realistic? What kind of ecological problems are likely to be encountered with all this degradable plastic floating around the countryside?

SCOTT. On the question of the amount of material which could be
used in mulching films, this depends of course on finding a successful system first of all. The dominant reason for changing to photodegradable mulching is economic. In other words, it saves the
cost of collecting up, washing, or even disposing of normal mulching materials so that it is difficult to extrapolate because this depends on the size of the economic gain. I know that in certain
countries, particularly in southern Italy, a very large amount of
mulching film is already used which is allowed to remain on the
soil and is not collected. In Israel where we have recently done
trials on degradable mulching film, they do collect it and it is
there that there is a great economic incentive to change over to
photodegradable materials. A trial was done last summer in
Israel and the only difficulty was that they found it photodegraded
a little bit too fast for them, but this is something we can cope
with quite easily by changing the concentration of the additive.
This is probably the only area where there is a positive economic

incentive to change over to degradable plastics. I am sure that
this will make it go, but it is almost impossible I think to assess
the size of the market.

As to the ecological effects of the product, we have done work
now on the low density and high density poly(ethylenes) in which we
have studied the biological breakdown, and judging by the trials
which have been done in Israel this summer this material goes to
a powder form within three or four months of outdoor exposure.
At this point it is a water wettable material which becomes readily
absorbed by the soil and ultimately when the molecular weight is
reduced to the order of 5000 it does biodegrade. There is no fore-
seeable problem even if you put this mulching film on the land every
year with an accumulation of this material in the long term. It is
claimed that it acts as a soil conditioner, but I thought I would
leave that for the commercial people.

QUESTION. Would the EPA gentlemen agree with that statement;
that there would be no ecological problem whatsoever to have all
this degradable polymer lying around the land?

ANSWER. I think the EPA answer would depend on some trial work
to see what the bugs are. Why try to come up with an answer on
pure supposition when, as you have heard, in Israel this is already
commercially being done and there are parts of this country that
have similar kinds of agriculture where I'm sure that it will be
looked at.

QUESTION. Can Professor Scott identify the chemical composition
of his preferred additive?

SCOTT. Do you want the chemical constitution? It is iron dibutyldithiocarbamate.

QUESTION. What are the ultimate degradation products of photodegradable plastics?

SCOTT. Obviously we haven't had the time to do this in the environment. What we have done is accelerate the tests by thermal and
oxidative degradation. This is in the absence of bacteria and

environmental agents. By extrapolation to room temperature or the kind of temperatures you find in the environment, we would think that for a normal oxidative degradation which is distinct from the biological degradation which takes over, the material would be reduced to long chain carboxylic acids and hydroxy carboxylic acids over a period of five years. But of course once you get to this stage of having carboxylic acids which are soluble or solubilized by the function groups then bacteria and particularly fungi can feed on it. This is what we have found, that these materials are able to support bacterial or fungal growth.

GUILLET. The question is difficult to answer only in this respect: in an exterior degradation you have a wide variety of different conditions. It won't be the same in the desert as it would be in a forest or on the beach so that times have to be given in orders of magnitude. We have done tests on a number of our plastics and the most successful of these, that is the fastest system, is propylene (ECOLYTE P). We have not identified the actual composition of all products. There are a lot of peaks on a gas chromatograph and identifying all of them would be difficult, but we do know enough about the chemistry to predict that we are going to end up with a lot of long chain branched carboxylic acids, esters and, to a lesser extent, ketones which come as a result of the decomposition of hydroperoxides. These are rather similar to the kinds of degradation products that you would get from many types of organic material of natural origin. In the case of both poly(propylene) and poly(ethylene), where this structure is observed in the degraded products, our tests in the soil for the biological oxygen demand in soil samples, indicates that we get about 10% degradation in a month. Now that's a very special situation where we prepared samples and mixed them with the soil carefully. In one case, a very rapid case, you would have complete decomposition in maybe a year or two.

QUESTION. How rapidly does biodegradation occur? Are the rates similar to those of, say, detergents?

GUILLET. People try to compare biodegradable detergents and biodegradable plastics. The rates differ by several orders of magnitude. With detergents you want the biodegradation to take place within a matter of a few hours of reaching the water supply. Something which lasts longer than that is a real problem in a municipal waste water sewage system. With plastics, once the material has broken down into particulate matter the litter aspect, which is essentially an aesthetic problem, has disappeared. Then it really doesn't matter whether it takes one year or five years or twenty years for the carbon to return to the biological cycle.

QUESTION. What are the reactions of consumer groups to these new developments? Will the housewives buy products in degradable plastic packages?

GUILLET. I have on file about 100 letters from various groups and consumer organizations who ask me to tell them where they can buy ECOLYTE so that they can replace other non-degradable materials from the supermarkets.

SCOTT. I would like to reinforce that very strongly. In fact, in the early days when our discoveries were announced, about 90% of the correspondence was from the United States. We were not too interested in the U.K., but the U.S. is certainly extremely interested. The women's institutes in the U.K have certainly taken it up and so have some of the pressure groups. The interesting thing is that the nearer you get to the user of the plastics, the stronger is the pressure and this pressure is gradually being brought back from the user of plastic packaging, in our country the big supermarket stores, to the plastics' fabricators, to the polymer manufacturers. In the U.K. at the moment, it has got as far as the plastics fabricators, but the polymer manufacturers have not yet really admitted that the problem exists.

WILES. We get letters too and there are a number of people that are interested in ecology that have been looking for schemes such as this and I think that through the normal political process in this country that, as availability of such materials becomes known, these groups will be pressuring the suppliers, the supermarkets for the increased availability of this material.

QUESTION. Will these new photodegradable plastics cause difficulties in waste or garbage disposal?

SCOTT. As I understand it, the question is can what we might call these ecologically-desirable plastics fit into a normal waste disposal system. The answer insofar as the system we have developed is concerned is yes. If the plastics are incorporated in a landfill, they behave entirely as normal plastic materials do if they are buried beneath the level where there is oxygen access. If, on the other hand, they are on the surface, where there is oxygen access, then they will degrade in the way that we have been discussing. They will oxidize and biodegrade and become part of a useful environment. So far as recycling is concerned, which is perhaps the most important system one can envisage for the future, by taking suitable action during the recycling process they can be recycled. In incineration they have no effect whatsoever. They don't interfere with the normal incineration process.

QUESTION. If you use water to scrub incinerator gases do you not replace an air pollution problem with a water pollution problem?

ENGDAHL. I think at the outset we should begin to look at what effects these new trends may have. A number of the constituents of flue gases, whether it be HCl from increased PVC, or HBr from addition of flame retardants, should be scrubbable so I think the best method now is to remove them via a water stream. Making them water-borne is another problem, but it can be solved.

QUESTION. Would it not be better to control the sources of ignition rather than reduce the flammability of materials by adding potentially toxic chemicals to plastics?

ENGDAHL. There is at least one chemical company which shall not be named which is actively at work on this problem but it is not an easy one. When you think of the immense research investigation and investment to control flammability, you realize that very little is being done to control the sources of ignition; 45% of all fires are from smoking! There's an imbalance in our research and development activities.

QUESTION. What is the likely increase in cost for photodegradable plastics?

SCOTT. Do you want me to speak as a scientist or a commercial man? Including royalty payments, I understand this would put onto the price in terms of raw materials cost less than 1% on the cost of the product. That is the cost of the resin would be increased by less than 1%.

GUILLET. I think the actual increase in cost of the chemicals involved is small in both processes and we expect that there should be no more than say a 5% increase in the cost to the ultimate consumer.

QUESTION. Is the rate of photodegradation affected by the thickness of the sample?

GUILLET. Twenty to 30 mils is not much of a problem in the styrene system that I was discussing yesterday. We would envisage even thicker specimens in foamed products since the degradation is taking place outdoors where there is a natural abrasion. As the poly(styrene) degrades in foam samples, the surface simply washes off. If you have a thick sample, it gets thinner and thinner and ultimately it gets to wafer size where it breaks up into small particles. There is a real problem with really thick solid samples, say one-half inch. On the other hand, we're not likely to have many packages made of specimens having thick cross sections. That's more used in an engineering plastic application.

SCOTT. We found some other unexpected effects on sample thickness. It depends to a very large extent on whether the material is pigmented and the type of pigment that is used. It also depends on the morphology of the polymer, whether it is largely an amorphous polymer or a crystalline polymer. For example, in low density poly(ethylene), in general the thicker the sample, the slower the rate of breakdown. But in high density poly(ethylene) the reverse is true. This is almost certainly associated with the crystalline content and it seems to be due to a preferential surface degradation which introduces cracking or micro cracks, which therefore lead to very rapid propagation. Up to a thickness of somewhere in the region of 240 microns, then the thicker the sample, the faster the rate of degradation. But once you introduce pigments, especially the darker colored pigments, then they act as very effective screening agents. At thicknesses as thick as one-eighth inch there would be a considerable delay time before breakdown occurs. We are trying to persuade the people in the U.K. who are looking at detergent containers, for example, to go to the thinnest-walled containers they can, particularly in high density poly(ethylene) and this seems to be providing a useful answer. The high density poly(ethylene) detergent containers can be made to go quite rapidly.

GUILLET. Not all of the problems can be solved by changing the resin. Another important aspect is the design of the package. We don't suggest that we can solve all problems by simply taking existing packages and making them out of these new materials. We have new materials which have certain properties and if you wish to have a disposable package, it is necessary to design the package around these materials. So you obviously don't put in pigments that absorb all the light, nor do you make a large thick section which won't degrade. Cooperation is necessary between the package designer, the material supplier and the man who is going to use the package and put his advertising on it. For example, you don't want to cover the package with black letters which won't degrade.

QUESTION. What will be the effect of large accumulations of degraded plastic in the soil?

SCOTT. Supposing that this soil conditioner that we're talking about accumulates, does this matter? Is it going to create any problem in the environment? It certainly won't create an aesthetic one because it won't be observable. Certainly as I see it, it is very similar to introducing humus material into the environment. Cellulosic material doesn't break down all that rapidly. It becomes accumulated in the soil and breaks down over a long period of time. I can't see that there is any real difference here. You know that the ultimate products, long chain carboxylic acids, will biodegrade. But if they are inert in the meantime, I can't see that they can possibly do any harm in the environment. Perhaps I am missing something here but we're talking here about materials which contain carbon and hydrogen. These are the only two elements they contain. The oxygen comes from the air.

QUESTION. What about PVC? Doesn't it contain chlorine?

SCOTT. This we did mention yesterday when we talked about PVC. We said that we were doubtful about the validity of breaking down high molecular weight chlorinated material which is inert, into low molecular weight clorinated hydrocarbons which are very far from inert. Now poly(ethylene) is a different matter. It contains only carbon and hydrogen. After oxidation it is very similar to cellulose.

QUESTION. What about the possibility that special micro-organisms will develop in garbage dumps which might destroy all plastics?

WILES. I take it that you are concerned about selectively breed-ing and collecting particular fungi and other flora in garbage dumps. This is a perfectly normal situation in the soil of a garbage dump, which you may wish to call a "bacteriological bank" in the sense that there are tens of thousands of fungi alone that are potentially viable in any piece of arable land. You create conditions which favor some over others. It is a perfectly normal situation. I am not quite sure of the cause of your concern.

QUESTION. Surely we need to know more about the long term effects of burying plastics in landfill areas.

WILES. A lot of this work is going on but I think you are, perhaps without knowing it, making a plea to government agencies to fund more fundamental R & D in this area. I would like to mention that most micro-organisms do have enzymes which reduce long chain carboxylic acids to acetic acid. This is similar to lipid metabolism about which a great deal is already known.

QUESTION. Would it not be better to use paper and other natural materials for packages? Making plastics degradable only makes the present situation worse by extending the use of synthetic mate-rials by improving their properties.

SCOTT. This is about the best advertisement that degradable plastics have yet had because the questioner is really saying that

degradable plastics offer the possibility of producing a non-damaging material in the environment, non-damaging to the ecology and
non-damaging to people. Glass and metal cut people! After all,
we have glass and metals now. The argument surely is to change
over to something which doesn't litter indefinitely. Let's try to
restore the balance to where we were with cellulosic materials.
This is all we are trying to do. In the process we are also introducing materials which are less damaging than glass and metal.

QUESTION. Surely the only way to control litter is through legislation. We must get rid of litter-bugs and spend more money clean-
ing up the streets!

GUILLET. Let me make a comment because the speaker has
brought up a common misconception. Those who advocate the use
of controlled-lifetime plastics do not suggest that we should immediately stop sweeping our streets, stop picking up litter where it
occurs and wait until litter degrades. This is not at all the point.
What we are saying is that there are large areas of the country
where litter exists where it is either not practical or desirable
to have crews going around to clean it up. It is obvious that the
fastest way to get rid of litter in cities is to pick it up.

On the deserted areas of the Caribbean where I go to get
away from it all, I find that the beaches are covered with plastic,
metal and glass containers which came from somewhere else.
Now if I am sitting there on the beach, I don't want to have a big
crew of men coming through the place picking up litter. There
are also many areas of the country in the north of Canada where
we go to get away from civilization, and we find a nice camping
place and it is covered with bottles and cans which have been there
for 20 years. I'm not sure that roadside litter is a good example
to use because you do have the possibility on a roadside of having
a crew go along and pick it up. We feel that as a matter of principle
you should not package an impermanent object in a permanent
package. It just doesn't make sense to wrap a sandwich, for
example, in a package that will last 50 years.

QUESTION. I don't think degradable plastics are the answer.
Surely the solution to the litter problem is better education. We
need to convince people that they should clean up after themselves!

GUILLET. I think the misapprehension is that you are suggesting we think we have "the solution" to the litter problem. We don't have "the solution". There is no one solution to the litter problem. But there are things that you can do to make the litter problem less acute, and that's what we propose. Make containers from degradble material if possible. Initiate legislative and educational programs. Provide incentives for recycling. All of these will help the situation.

QUESTION. Won't degradable plastics cause problems of biological oxygen demand in lakes and rivers?

SCOTT. I think you may be referring to a remark I made yesterday when I said I didn't think these photodegradable plastics would be a problem in water supplies. The time scale of the biodegradation process is so long that they would be deposited from the water supply from a river for example, long before any accute biological demand occurred. It is unlikely that the biological demand will be acute because the process is so slow. It will not make any difference to the water situation because these materials are deposited from water. They are not like detergents which are soluble in the water and must remain there. These plastic degradation products are deposited from the water. Although they are water wetable, they are nevertheless not soluble, and they are not carried indefinitely in the water.

QUESTION. Will the U. S. Government require environmental impact statements soon for all packaging materials? How long will it take to get approval for these new materials?

INGLE. It is going to take the Internal Revenue Service probably more than a year if it does succeed in providing a convincing environmental impact statement for PVC rigid bottles for alcohol. Now if the EPA does have its own way and does get the power to require environmental impact statements for food and packaging materials, I can see on the basis of some of the arguments which have been raised this morning that there will be a substantially longer period in proving the environmental impact to be negligible for this type of material.

QUESTION. Professor Scott, does your process work with poly-(styrene)?

SCOTT. The real answer to this is that we've done far less work on poly(styrene) than we have on the polyolefins. We wanted to go for the bulk plastics first. I think if you are thinking of poly-(styrene), Professor Guillet's type of solution is probably more immediately applicable. The process which we have developed does certainly work with poly(styrene), but we have not advanced nearly as far as Professor Guillet has.

QUESTION. What about the rate of photodegradation in styrene foam compared to sheet?

SCOTT. I don't think there is any difference in the relative rates of the photodegradation process. We get the same sort of order of acceleration. One of our problems is that we haven't in my organization got facilities for blowing the poly(styrene) foam, which is what we would have liked to have done. This is a major problem, of course, foamed plastic that's lying around. I think it certainly will work, but Professor Guillet has obviously got a lot further than I have at the moment.

QUESTION. Are the new nitrile-based plastics degradable?

SCOTT. Not in our experience. We've done comparisons of these nitrile-based packaging materials with the polyolefins, for example, and I would say that the nitrile-based materials are much more similar to PVC than they are to the polyolefins in that they last very much longer. I would not by any means call the nitrile-based materials photosensitive. They can be made photosensitive by the type of process which we have been talking about, but that's a different matter. They are not themselves photodegradable.

APPENDIX

PROGRAM

Symposium on Polymers and Ecological Problems

Georgian Room, Statler Hilton Hotel
New York, New York

General Chairman: James Guillet

Monday, August 28, 1972
Chairman and Presiding Officer: D. M. Wiles

9:00 AM	Chairman's Introduction. J. Guillet
9:10	Photostabilization and Photodegradation of Plastics. B. Baum, R. A. White
9:40	Discussion
9:50	A Delayed Action Photo-Activator for the Environmental Degradation of Plastics. G. Scott
10:30	Discussion
10:40	Polymers with Controlled Lifetimes. J. Guillet
11:20	Discussion
11:30	The Biodegradability of Synthetic Polymers. J. E. Potts, R. A. Clendinning, W. B. Ackart, W. D. Niegisch
11:55	Discussion

Monday, August 28, 1972
Chairman and Presiding Officer: W. J. Bailey

2:00 PM   Chairman's Introduction. W. J. Bailey

2:10   An Industry View of Plastics in the Environment.
               G. W. Ingle

2:40   Discussion

2:50   The Plastics Issue. G. E. Huffman, D. J. Keller

3:20   Discussion

3:30   Federal Guidelines for Recycling of Polymers.
               J. P. Lehman

4:00   Discussion

4:10   Legislative Problems with Plastics. R. Black

4:40   General Discussion

Tuesday, August 29, 1972
Chairman and Presiding Officer: R. F. Schwenker, Jr.

9:00 AM   Chairman's Introduction. R. F. Schwenker, Jr.

9:10   Effluents from Incineration of Plastics. R. B.
               Engdahl, H. Krause, Jr., P. D. Miller

9:30   The Autoignition of Multicomponent Fiber Systems.
               B. Miller, J. R. Martin, C. H. Meiser, Jr.

10:00   Thermal Analysis of Irradiated Poly(Vinyl Chloride).
               R. Salovey, R. G. Badger

10:25   Recycling Poly(tetrafluoroethylene). B. Arkles

10:50   Break

11:00   Panel Discussion on Plastics and Ecological
               Problems. D. M. Wiles, Moderator